Emerging Trends
in
Teaching of Chemistry

KANISHKA PUBLISHERS, DISTRIBUTORS
4697/5-21A, Ansari Road, Daryaganj,
New Delhi-110 002
Phones : 2327 0497, 2328 8285
Fax : 011-2328 8285
E-mail : kanishka_publishing@yahoo.co.in

Emerging Trends in Teaching of Chemistry

First Published 1996

Second Edition 2007

ISBN 81-7391-040-5

PRINTED IN INDIA

Published by Madan Sachdeva for Kanishka Publishers, Distributors, 4697/5-21A, Ansari Road, Daryaganj, New Delhi-110 002. Typeset by Sunshine Graphics, Delhi, and Printed at Chaudhary Offset Process, Delhi.

Preface

Science teaching has some specific purposes besides the general purposes of education which are as following:

1. An understanding of and the ability to use those fundamental concepts and relationships of science that enable the individual to interpret intelligently the conditions and happenings in the natural and the man-made environment.
2. A knowledge of and the ability to select and use scientific facts and information that help the individual to know himself and the materials and forces of the environment.
3. Skills of observation, reflective thinking and problem solving which enable an individual to face questions encountered in living and to approach answers to such questions.
4. Mental and manipulative skills which make the individual able to find success and enjoyment in using materials and forces to meet both personal and social needs.
5. Attitudes, standards and ideals relating to the materials and forces of the Universe that favour an individual's living with appreciation, poise and success in the midst of a complex environment.
6. Attitudes, standards, and ideals relating to scientists and citizens both past and present that will help to guide and guard and individual in his living with his associates and with other people. The teaching of various disciplines of science broadly correspond to the above objectives. These aims of teaching science hold good both in the developed and the developing world. The present volume in an exercise in this direction.

The present publication is based on current syllabus and both traditional and modern methods of the teaching of chemistry. The structured contents include; a focussed comment on the science classroom and the ways of the scientists; approach to teaching the ways of the scientist; science teachers and the chemistry course contents; secondary school chemistry and school science facilities; basic philosophy of teaching chemistry; the important units of—man and the nature of matter, the structure of matter, types of chemical change, and water, solutions and non-solutions, with reference to acids, bases and salts—the metals and their compounds, carbon and its compounds, and useful non-metals. It is earnestly hoped that the present publication will be an indispensable aid to the teaching of chemistry for all teachers.

T.N. Ratho

Ravi Prakash

Contents

1

The Science Classroom: Ways of the Scientist

A note at the beginning. We have at hand 12 syllabuses, from 37 states, for general science, biology, physics, chemistry, earth science, and physical science courses. They have one thing in common: all propose to teach *scientific method.* Forty one of them seem to deal with the "empirical approach the slowest, least effective way of "problem solving."

Actually the word "science" stands for such a complex variety of information, abilities, and operations that none of the many published definitions seems wholly adequate. We hesitate to add one more effort to compress the grandeur of science and scientific work into a brief pattern of words. However, perhaps we can clarify, what science *isn't*, and suggest explicity what it *involves.* Many eminent scientists and philosophers have written about the nature of science; those readers who wish to go beyond our discussion may find the books listed at the end of the chapter helpful.

One peculiarity of ours will certainly not escape you: we tend to think of science more as a *verb* than as a *noun*; we tend to think of the way a scientist *works* rather than what science *is.* We think that science is more concerned with the process by which a body of reliable knowledge is obtained than with the resulting body of knowledge itself. Consequently when we talk of *science*, we shall really be taking about *ways in which scientists seek concepts.*

The So-Called "Scientific Method"

To predict what an individual scientist will do is hazardous, for as individuals they do a variety of activities in different ways. However, then are some common attributes of "scientific" behaviour, and these are what we are searching for. Unfortunately a great myth has developed that the scientist has some special, even regular, way of proceeding; he has, too many people believe, a "scientific method," which is the key to his success. Often this scientific method is reduced to five steps:

1. Define a problem.
2. Gather relevant data.
3. Form hypotheses.
4. Test the hypotheses.
5. Reach a conclusion.

No one will deny that scientists do all these things, but knowing that they "define problems" is no more help to us in doing likewise, or even in understanding what they are doing than knowing that artists "paint pictures." What this scheme does not tell us is exactly what we need to know; What operations are represented by the verbs? Let us describe a "method" of mountain climbing in the same way:

1. Select an unclimbed mountain.
2. Organize a party.
3. Plan the equipment.
4. Go to and climb the mountain.
5. Plant a flag on the top.

Mountain climbere do these things: but does this analysis

help you to "select", "organize," "plan", or "climb"? The analysis may be correct but it is sterile; it does not aid us in performing or even understanding these operations.

Percy Bridgman, Nobel Laureate in Physics, cuts a broad swath through the definition of the methods of science by a general statement which appeals to many other scientists: "The scientific method, as far as it is a method, is nothing more than doing one's damnedest with one's mind, no holds barred."[1] He goes further, in an article addressed to teachers, to say (the italics in the final paragraph are ours):[2]

> It seems to me that there is a good deal of ballyhoo about scientific method. I venture to think that the people who talk most about it are the people who do least about it. Scientific method is what working scientists do, not what other people or even they themselves may ask about it. No working scientist, when he plans an experiment in his laboratory, asks himself whether he is being properly scientific, nor is he interested in whatever method he may be using as *method.* When the scientist ventures to criticize the worh of his fellow scientist, as is not uncommon, he does not base his criticism on such glitrering generalities as failure to follow the "scientific method", but his criticism is specific, based on some feature characteristic of the particular situation. The working scientist is always too much concerned with getting down to brass tacks to be willing to spend his time on generalities.
>
> Scientific method is something talked about by people standing on the outside and wondering how the scientist manages to do it. These people have been able to uncover various generalities applicable to at least most of what the scientist does, but it seems to me that these generalities are not very profound, and could have been anticipated by anyone who knew enough about scientists to know what is their primary objective. I think that the objectives of all scientists have this is common—that they are all trying to get the correct answer to the particular problem in hand.

This may be expressed in more pretentious language as the pursuit of truth. Now if the answer to the problem is correct, there must be some way of knowing and proving that it is correct; the very meaning of truth implies the possibility of checking or verification. Hence the necessity for checking his results always inheres in what the scientist does. Furthermore, this checking must be exhaustive, for the truth of a general proposition may be disproved by a single exceptional case. A long experience has shown the scientist that various things are inimical to getting the correct answer. He has found that it is not sufficient to trust the word of his neighbour, but that if he wants to be sure, he must be able to check a result for himself. Hence the scientist is the enemy of all authoritarianism. Furthermore, he finds that be often makes mistakes himself and he must learn how to guard against them. He cannot permit himself any preconception as to what sort of results he will get nor must he allow himself to be influenced by wishful thinking or any personal bias. All these things together give that "objectivity" to science which is often thought to be essence of the scientific method.

But to the working scientist adisell all this appears obvious and true. *What appears to him as the evence of the situation is that he is not consciously following any prescribed course of action, but feels complete freedom to utilize any method or device whatever, which in the particular situatian before him seems likely to yield the correct answer. In his attack on his specific problem he suffers no inhibitions or precedent or authority, but he is completely free to adopt any course that his ingenuity is capable of suggesting to him. No one standing on the outside can predict what the individual scientist will do or what method he will follow. In short, science is what scientists do, and there are as many scientific methods as there are individual scientists.*

Science progresses as the explanations (or concepts)[3] it produces for preciously unexplained things or events both fit the facts on which they were based and lead to new predictions.

which can be tested. The results of these tests may or may not agree with these predictions. If they to agree, our confidence in the explanation is bolstered, for we then have added one more set of phenomena which we can predict. If, however, after repeated experimental work our expectations and observations do not agree, or do not agree as well as we would like, then our explanation must be re-examined to remove, if possible, the new discrepancy. And so we have a continuous round of discrepancy, explanation, prediction, test, new facts, new explanation.

The self-correcting process is basic to all scientific work. But individual scientists, despite their best intentions, are biased, misinterpret data, and formulate erroneous explanations. Then how is it that we attain any increase in predictive ability? Independence, in science at least, has its price; the independence of each individual, as Bridgman noted, imposes upon him a continuous attitude of constructive skepticism toward his own work and that of others. Diversity of explanation of known data is not only unavoidable but essential: indeed, diverse explanations are assiduously sought. A fundamental aspect of science, as it has evolved over the centuries, is that the alternate answers which arise in this way, incomplete or erroneous as they probably are, are so formulated that each is susceptible to additional testing, and thus is productive of further concept attainment.[4] Concept attainment is a goal of the scientist.

Consider, for example, the explanation of the origin of continental glaciers, the great ice sheets which a few thousand years ago covered a considerable part of the United States. Evidence convinces us that such ice sheets did exist. But what was the mechanism of their formation? Can we predict the conditions under which they would reappear? Willlam L. Stokes stated that to his knowledge twenty-nine different explanations were in the literature, and he went on to say that "Most of these had little chance of survival from the first, but others enjoyed some degree of success until they were rendered untenable by subsequent accumulated information." Not content, he went on to propose a thirtieth explanation.[5]

Such a multiplicity of effort may seem wasteful, but it surely is not; only by having a variety of different trial explanations, based on different selections of data and physical mechanisms, can be obtain the clues we need in order to make further useful observations. These observations can be expected to weed out many of the trial explanations.

Empiricism—One Way of the Scientist

Many discussions of science, or of "scientific method", point to empiricism as the essence of science, as do the syllabuses mentioned at the beginning of this chapter. Let us take a look, for a moment, at this empiricism and see what it is.

Empiricism and Common Sense

Conant[6] has illustrated the sence of empiricism, the "cut and try" approach, by a commonplace example. Take out a set of keys and ask someone to unlock a given door. Clearly the approach to the solution of this problem of empirical; lacking any clues, any basis of selection, any hypothesis to guide him, the "experimenter" will *try* the keys until he finds one which opens the door. This follows the route of the systematic "try it and see" or "trial and error" approach. The "method of solution" might even be placed on the blackboard thus:

Problem: To open the door

Apparatus: Keys

Method (*observations*: Try in sequence each key

Operation: Systematic trials

Conclusion: Key No, 1007 opens the door

If the reader objects and says that this is simply common sense, he is correct. The "try if and see" (empirical) approach is a common sense approach, and it is used in many fields of endeavour besides science. Much of the laboratory work useful in defining the characteristics of new phenomena is precisely this a systematic "try it and see," or *empirical* approach. Note that empiricism when effective is *systematic*: that is, it is guided by ideals and concepts. Without such guiding ideas, activity in concept attainment is futil.

Empiricism in the Laboratory

Let us book at a laboratory investigation which exhibited a high degree of empiricism: the making of complex iron alloys. Small changes in the percentages of many elements (carbon, silicon, phosphorous, nakel, or sulphur) were known to alter the properties of the final allow. Yet desired properties could not be obtained by prediction: for no reliable generalisms explana-nation, no *concept* about the interactions of several metallic components, had beea proposed. The systematic making and testing of many individual samples was the only way to discover how they would behave. Later, desired, any of the sample alloys could be duplicated, but the properties of a new alloys could not be predicted, not could an alloys of given properties be designed.

This is an example of almost pure empiricism: empirical allows us to choose only among the things we have already done. Thus empiricism though indispensable, when unaccom-panied by some effort to attain *prediction* is a slow and fairly unsatisfactory way to get results.

Empiricism in the Classroom

Crude empiricism often appears in the classroom. A question is asked, by the teacher or a student, which leads possibly to one of the following statements below:

Let's do an experiment.

We're scientists. Let's *try it and see*?

Suppose the question is "Does air have pressure?" (It is desired to believe that students would worry very much about this, but the human remains a standard one). The teacher writes the problem on the board; "Does air have pressure?" Then an outline on the board follows, something like this:[8]

Problem: Does air have pressure?

Apparatus needed: A can, burner, stopper, tripod, and water.

Observations: (Written by students in their notebooks. This is a description of what they "tried" and "saw,") "We heated the water in the can until stream appeared. Then we put a stopper in the can. As the can cooled, its walls began to collapse."

Argument: Since the steam displaced the air, a partial vacuum was produced inside. The collapse of the can must have been due to a force acting on the outside. *This force is the pressure of air*. (No other explanation seems adequate.)

Conclusion: Air exerts pressure.

There is nothing wrong with using simple empiricism; we all do is daily and the scientist does it too. However, in the example just cited a great many things are wrong, both scientifically and pedagogically. No real problem existed for the students; for there was no discordant observation, no trial explanation, not even an "educated guess." There was no testing of a hypothesis by testing its consequences. There was no design or even selection of equipment with which to gather the needed information. And, of course, there could be no reasoning to fit the new information to an initial explanation.

This demonstration, done to the death, is really quite sophisticated. The student cannot see what is happening in the can, and he must recall considerable information if he is to "believe" the explanation desired. His tacit agreement to many concepts (principles) is assumed: water is vapour occupies a much larger space than as liquid: water vapour will displace air; upon cooling water vapour condenses with a lowering of the gas pressure (inside the stoppered can). Then he may be able to visualize the difference in pressure inside and outside the can which resulted in its collapse. Finally attention is focused on the question: What is the origin of the external pressure? In short, more than the empirical approach is necessary to understand what is really happening—insight, for instance, as well as prior experience and reading.

Ways of the Scientist: Planned Investigation

So far we have attempted to describe empiricism, the "try it and see approach. We have seen that it is indeed a part of the scientific approach; we cannot picture a scientist unwilling to go into the laboratory and "to it." But "it" is always a specific operation or reaction seemingly worth the effort to "try." In other words, more than empiricism is involved in science.

Let us examine a classic investigation and analyze the operations involved, both physical and intellectual, from beginning to end.[9]

Redi's Investigation[10]

It being thus, as I have said, the dictum of ancients and moderns, and the popular belief that the putrescence of a dead body, or the filth of any sort of decayed matter engenders worms; and being desirous of tracing the truth in the case, I made the following experiment;

At the beginning of June I ordered to be killed three snakes, the kind called eels of Aesulapius. As soon as they were dead, I placed them in an open box to decay. Not long afterwards I saw that they were covered with worms of a conical shape and apparently without legs. These worms were intent on devouring the meat, increasing meanwhile in size, and from day. to day I observed that they likewise increased in number; but although of the same shape, they differed in size, having been born on different days.

But all, little and big, after having consumed the

meat, leaving only the bones intact, escaped from a small aperture in the closed box, and I was unable to discover their hiding place. Being curious therefore, to know their fate, I again prepared there of the same snakes, which in three days were covered with small worms. These increased daily in number and size remaining alike in form, though not in colour. Of these, the largest were white outside, and the smallest ones, pink. When the meat was all consumed, the worms eagerly sought an exit, but I had closed every aperture. On the nine. teenth day of the same month some of the worms ceased all movements, as if they were asleep, and appeared to shrink and gradually to assume a shape like an egg. On the twentieth day all the worms had assumed the egg shape, and had taken on a golden white colour, turning to red, which some darkened, becoming almost black. At this point the red, as well as the black ones, changed from soft to hard, resembling somewhat those chrysalids formed by caterpillars, silkworms, and similar insects. My curiosity being thus aroused, I noticed that there was some difference in shape between the red and the black eggs [pupae], though it was dear that all were formed alike of many rings joined together; nevertheless, these rings were more sharply outlined. and more apparent in the black than in the red, which last were almost smooth and without a slight depression at one end, like that in a lemon picked from its stalk, which further distinguished the black egg like balls. I placed these balls separately in grass vessels, well-covered with paper, and at the end of eight days, every shell of the red balls was broken, and from each cease forth a fly of gray colour, torpid and dull, misshapen as if half finished, with closed wings but after a few minutes they commenced to unfold and to expand in exact proportion to the tiny body which also in the meantime had acquired synnetry in all its parts. Then the whole creature as if made ahead having best its gray motor, rock me a most brilliant and vivid green; and the whole body that expanded and grown so that it scerned incredible that it could ever have been contained in the small shell. Though the red eggs [pupae] brought form green fiels at the end of eight days, the black ones laboured fourteen days to produce certain large black files striped with white, having a hairy abdomen, of the kind that we see daily buzzing about the butchets stails.

Having considered these things, I began to believe that all worms found in meat were derived directly from the droppings of flies, and not from the putrefaction of the meat, and I was still more confirmed in this belief by having observed that, before the meat grew wormy, flies had hovered over it. of the same kind as those that later bred in it. Belief would be in vain without the confirmation of experiment, hence the middle of July, I put a snake, some fish some eels of the Arno, and a slice of milk-fed veal in four large, wide-mouth flasks; having well closed and scaled them, I then filled the same number of flasks in the same way, only leaving these open. It was not long before the meat and the fish, in the second vessels, became wormy and flies were seen entering and leaving at will: in the closed flasks I did not see a worm, though many days had passed since the dead flesh had been put in them. Outside on the paper cover there was now and then a deposit, or a maggot that eagerly sought some crevice by which to enter and obtain nourishment. Meanwhile the different things placed in the flasks had become putrid.

Not content with these experiments, I tried many others at different seasons, using different vessels. In order to leave nothing undone. I even had pieces of meat put under ground, but though remaining buried for weeks. they never bred worms. as was always the case when flies had been allowed to light on the meat. One day a large number of worms, which had bred in some buffalo meat, were killed by my order; having placed part in a closed dish, and part in an open one, nothing appeared in the first dish, but in the second worms had hatched, which changing as usual into egg-shaped balls [pupae], finally became flies of the common kind in the same experiment tried with dead flies, I newer saw anything breed in the closed vessel.

Leaving this long digression and returning to my argument, it is necessary to tell you that although I thought I had proved that the flesh of dead animals could not engender worms unless the semina of live ones were deposited therein, still, in order to remove all doubt, as the trial had been made with closed vessels into which the air could not penetrate or circulate, I wished to attempt a new experiment by putting meat and fish in a large vase closed only with a fine Naples veil that allowed the air to enter. For further protection against flies, I placed the vessel in a frame covered with the same net. I never saw any worms in the meat, though many were to be seen moving about on the net-covered frame. These, attracted by the odour of the meat, succeeded at last in penetrating the fine meshes and would have entered the vase had I not speedily removed them. It was interesting, in the meanwhile' to notice the number of flies buzzing about which, every now and then, would light on the side net and deposit worms there, I noted that some left six or seven at a time there and others dropped them in the air before reaching the net. Perhaps these were of the same breed mentioned by Scaliger, in whose hand, by a lucky accident, a large fly deposited some small worms, whence he drew the conclusion that all flies bring forth live worms directly and not eggs. But what I have already said on the subject proves how much this learned man was in error. It is true that some kinds of flies bring forth live worms and some others eggs, as I have proved by experiment.

Radi "tried it and saw." However, this is not all he did: let us analyze his approach. Our two-columns from separates the "evidence" (What Redi Said) from our generalized comment.

Radi's "Pattern of Investigation"

1. [There is a] dictum that the putrescence of a dead body, or the fifth of any sort of decayed matter, engenders worms; and being desirous of tracing the truth. . . made the . . . experiment. (lines 1-4)	1. A widely accepted explanation has never been tested. Suspicion is aroused. A test is designed.
2. Having considered these things I began to believe that all worms found in meat were derived directly from the droppings of flies, and not from the putrefaction of the meat. (Lines 46-48)	2. The preliminary test has yielded observations which do not agree with the accepted explanation. These observations suggest an informed guess—a working hypothesis. A new concept has been formed; it deserves testing.
3. Belief would be useless without the confirmation of experiment. hence in the middle of July, I put . , . (lines 50-54)	3. An experiment is designed to test this hypothesis under devised conditions. A control is used for comparison.

4. It was not long before the meat and fish in the [open] vessels became wormy . . . [but] in the closed flasks I did not see a worm. (lines 54-57)

4. The experimental design of the operation is found useful. Differences between the experiments and control cases support the tentative hypothesis; they do not support the accepted belief.

5. Not content with these experiments, I tried many others at different seasons, using different vessels. (lines 62-63)

5. Further observations are made in order to generalize the observation.

6. In order to remove all doubt . . . I prepared a new experiment . . . (lines 75 ff.)

6. A further modification in design broadens the evidence to clarify the effects of the air.

7. I never saw any worms in the meat . . . (lines 10 ff.)

7. The working hypothesis has been checked, minor variables have been eliminated, and the hypothesis has been supported by all the available evidence.

We may generalize this pattern taking into account the work of many scientists:

1. There seems to be a difference between what is observed and what is expected, or a vague explanation has been generally accepted without testing.
2. The observer analyzes the situation and recognizes a problem vague or distinct.
3. He makes one or more educated guesses as to the solution of the problem. This may be called a trial explanation or working hypothesis, limited or grand, depending on the magnitude of the idea. He may formulute several alternate hypotheses ("might be's"),

all equally consistent with the data available (induc, tion).

4. Each hypothesis is checked for consistency with facts already known through prior personal experience or accepted from the reports of others.

5. Any hypotheses which pass this test must then be tested further. Careful planning results in the design of experiments or planned observations (if..., then...) to test each of the hypotheses (deduction). Conditions are sought in which two hypotheses lead to differing predicted results; this speeds the sifting of hypotheses.

6. New evidence is acquired from these experiments or observations.

7. The "new concept" stated in the form of a hypothesis (or hypotheses) is either confirmed, discarded, or modified on the basis of the new evidence.

8. Any hypothesis (or hypotheses) which seems to fit the observations resulting from the experiment designed is accepted *tentatively* as a satisfactory answer to the question. The concept has been attained even though it will not necessarily be held permanently.

9. Many attempts are made to find exceptions to, or limitations of, the tentative conclusion. It is checked again and again in varying situations. Confirmation (or proof of error) by other scientists is sought.

10. If the conclusion continues to survive the work of other scientists, it is accepted as a useful addition to the knowledge with the field. It may still be discarded or modified whenever new evidence shows this to be necessary.

11. As more and more predictions from the hypothesis are confirmed, greater credence is assigned it. This is a gradual process, for the hypothesis always remains somewhat under suspicion. Concept attainment is never ending.

Once major clue to the scientist's approach is indicated in 8, 9, 10, and 11 above. It lies in a *deliberate attempt to defeat one's own conclusions. Above all, scientists want to know when they are wrong.* As a matter of fact, all results are stated in such a way that they *can be* proved wrong. It often takcs scientists a long time to develop the technology by which they can design an experiment which will test a hypothesis; but they can be sure that if a statement is made in such a way that a test of it is theoretically impossible, it is not in the area of science. For example, the hypothesis, "Empty space is full of a material with no observable properties," is meaningless; scientists try not to make such statements, which are impossible to prove or disprove, not for lack of the proper equipment, but simply because they never could bc tested, no matter what equipment was available. Niels Bohr emphasized this skeptical attitude; "Truth is something" that we can attempt to doubt, and then perhaps, after much exertion, discover that *part* af the doubt is unjustified."

Ways of the Scientist: Key Operations

We have seen from a study of Redi's work that there is sometimes a discernible pattern to a scientist's investigation, his way of getting at concepts. Is Redi's investigation a template of what all scientists do? If we examine the key operations of different scientists perhaps we can find some things common to them all.

Before the Problem

Any study begins with a familiarity with the phenomenon, with "what sort of thing it is." Otherwise one doesn't recognize a discrepancy, or a dissatisfaction with things as they arc. Concept seeking begins with the realization that the concept in hand is not adequate to fit the observed events.

This may be evidenced by a vague feeling of dissatisfaction with things as they are, or it may be as concrete as ',I wonder what is the pressure of the air on the top of the Washington Monument." The problem—something to be explored, to seek the answer to—must be interesting or tentalizing or "fun." (For students it should have meaning; it is expected that they will work hard. But the problem *should not be so complex* that there is little or no hope that the pupils, with the background and experience they have had, can organize a usable explanation, that is, attain the concept.)

Casual observation and description is followed by recognition of those aspects which begin to seem important. Such descriptions and classification are important in seeking any explanation, but they are part of the *beginning*, not the end, of the study. Carefully gathered data are the stuff of which explanations are built, but they may have, as Henri Poincare remarked, the same relation to a scientific concept as a pile of stones has to a house.

Note that we have said "carefully gathered data"; the data must be sought carefully and must be relevant to the problem. Simpson, Pittendrigh, and Tiffany[11] put it this way:

> Fact gathering in the hands of a scientist is a cultivated art; there is an infinity of facts in the world, and at any given time only a few of them are of vital importance to him.
>
> Are you being a scientist when you count the sand grain on Coney Island beach? No. It is true you are gathering facts . . . but you are probably crazy. Scientists gather facts that are *relevant* . . . to some theory to be tested or extended.

Hence the problem must be clearly stated, and, what is as important, the kind of explanation or solution suspected must be sketched in *advance* of experimental work. In fact, scientists "see" the solution almost at the same time that they "see" the problem. Otherwise, how can one decide what to do and what to expect, what is significant at any one time?

Classification

Scientists simplify their many observations of the world by grouping them according to many classification systems. Any classification system, of course, is valid only to the extent that it is useful. In botany, for example, various bases for organizing the plants of the world had been used before Linnaeus proposed using the characterestics of the reproductive organs for a new system. Botanical scientists have adopted this system, but practical gardeners may be more interested in the height, colour, and date of blooming of various plants, and a classification based on these observations might be valid for them. That is, each classification system is chosen to serve one end, and other systems of organizing the same material may be used for other ends. There is nothing sacred about a classification system; is purpose is to be useful in organizing one's observations.

Grouping observations is then a major part of scientific work. Often major new concepts have followed from the introduction of Novel and more useful ways of grouping data. Perhaps the most impressive example is the periodic table of the chemical elements. This patterning on the basis of limited chemical information later accommodated, even led to much of the information on the physical properties of atoms.

Generalizing from Data

Classification, then, seems to involve generalizing (or explaining) from data; this is essential to science, for it is one of the patterns of the investigator. But a scientist is necessarily skeptical, from bitter experience, of easy *over-generalization*, "for the truth of a general proposition may be disproved by a single exceptional case."[12] (We teachers are perhaps guiltier in this respect that scientists; in our mad haste to "cover" some field of science, we daily encourage dangerous overgeneralization.

Formulating and Eliminating Hypotheses

After the problem has been stated, simple guesses (or hunches of what sort of explanation might work are generally made. These hunches arises from similarities to past experience—the sort of explanations that have been useful for similar cases in the past. But a new problem is not likely to be answered by a simple copy of what worked in some other circumstances. Ingenuity and imagination in restructuring or regrouping the factors are important. (All too often the importance of this creative mental work is overlooked, for when "thinking" a student may not seem to be "doing" anything. This kind of inductive detective work toward a satisfactory explanation is the essence of science.

At least two important screens must operate in the sifting of possible explanations. First, any satisfactory explanation must account for a fair part of the available imformation. We do not say thal *all* of the available data must support the explanation, Sometimes more of the alleged data are incorrect; and we find that investigators, for their own peace of mind, often repeat experiments of others, especially when these seem to be of major importance. Second, some of the apparently relevant data may not actually be so.[13] Successful sifting and assaying of data, determining which of them *must* fit the hypothesis, is the mark of investigative talent.

Note that we have not yet placed the scientist in the laboratory (except perhaps to check some data which seemed dubious). He has been reading and thinking checking his hypotheses against Prior knowledge and experience to see if they are inconsistent with what is already known and then aginst the literature to see what other scientists have reported and proposed as explanations. Only the guesses which survive these tests need then be tested directly, that is, need be stated as problems which deserve investigntion. *And in this fact lies the main departure faom pure empiricism.*

Testing the Remaining Hypotheses

A scientist tries not to waste time: that is, he tends to try to work effectively and efficiently, He tries to eliminate possibilities as rapidly as possible: he uses *it then* reasoning. He takes each of his hypotheses and sees what it implies: "If this is true, *then* . . . must follow." For instance, if x germ causes tuberculosis, *then* the experimental animals injected with x germ should get tuberculosis, while the controls should not. He then tests these consequences by experiment or observation. (Formal logic tells us that when the "then" part of the statement, the implication, is false, the "if" part, the premise, must also be false.)

This process weeds out the alternate hypotheses rapidly. It may happen that the new observions eliminate *all* the alternatives we have; then major new ideas (concepts) are needed.

Before any scientist bothers others with his ideas and suggestions, he must carefully appraise them. He must be, as we have observed earlier, diligently self-critical. Only when he has tested his ideas again and again, and has checked them against the work of others, can he prepare the report required of him; this report is written rather like a legal brief, presenting has argument and citing supporting and limiting evidence from others Deway[14] long ago pointed out the great difference between the "dynamic logic" a scientist practiced in his laboratory and the "formal logic" he finally wrote in his papers. Too often we confuse the formal presentation with the hurly-burly of creative work. Too often we expect immature pupils just beginning to practice the rudiments of "dynamic" science to behave in the pattern of the the "formali" results we read.

Translating Ideas into Equipment

One major aspect of scientific work often overlooked is the ability to translate an idea into equipment. A wide knowledge of avilable instruments and their characteristics is essential: then a choice of components and arrangement must be made. Not everyone is equally effective at this so some scientists become specialists on instrumentation. Without carefal consideration of what questions we are asking with the equipment. we may get without realizing it, anwers to questions we did not intend to ask.[15]

The tools we use define the operation through which we gain information. While certain main-sized objects like trees and mountains may be examined as they are our knowledge of them may remain descriptive, superficial, and state. If we wish to explain in more detail what they are composed of or how they are changing, we dissect them physically and chemically. But then we are concerned with invisible materials and forces revealed to us only through gross effects upon our instruments. Louis Agassiz advised his students to "observe natures": but we must relize that for the study of invisible materials and forces, we can get answers only to the questions we ask through the arrangement of our instruments. Every arrangement of of apparatus has inherent within it the subtle story of "how we know what we know." Our knowledge of the structure of the atom has been influenced by the cyclotron; of the composition of the stars by the spectroscope.

Searching for Precision

All our knowledge of the world begins as qualitative information (description by adjective of adverb) and it is often vague at that. Comparisons are frequently made on qualitative bases.

In some instances we want to make nice, precise predictions, Then we try to replace the qualitative prediction about "what kind of . . ." with quantitative predictions about "how much of. . ." But the effort to make quantitative observations is great and is not always necessary for the prediction at hand. For example, for a storekeeper who wants a bright sign the distinction between argon and neon glow tubes can be made simply by noting that argon has a purplish light while neon is orangish (a qualitarive classification): he need not specify or measure the wave lengths of the individual radiations emitted (a quantitative classification). However for other uses it is important to know the wave lengths to a high degree of precision. Note, too, the deliberate distinction made between qualitative and quantitative chemistry. We must know that an objeect or reaction has certain qualitative characteristics before we can even choose whether or not to brother to determine them quantitatively.

We must always decide what degree of precision is useful for the problem at hand. There is no need to crack a walnut with a pile-driver, yet sometimes we fail to help students estimate what precision will be satisfactory for the problem being attacked. This point is exceedingly important, for as we all know, increased precision is obtained only with a severe price in time and equipment. Inevitably we are faced with the question of how exact our prdictions are and how exact our tests of them need be.

Accepting a Hypothesis as a Theory

When a hypothesis has been found which agrees with known data, is not inconsistent with other major concepts, and does not imply false consequences, what more should we ask of it? Just one thing: Does it work? That is does the new explanation lead to predictions that fit better with what is observed? This is what Conant meant when he defined science as "a series of conceptual schemes (theories) arising out of experiment or observation and leading to new experiments and observations."[16]

Sometimes a new explanation will be seen to account for peculiar observations which have been available but not explained previously. This often occurs after publication when specialists familiar with other sets of data suddenly see that the new idea accounts for previously unexplained observations in their field.

But above all scientists require that they know when their theories are wrong. Their theories, their explanations must be *testable*. Te repeat—does it work, where, when, and how well? Reflection on these brief comments about scientific work should remind us that science is built in the minds of men.

> [Natural science] expends its utmost pairs on attempting to describe the "how" fully and accurately by first-hand observations. . . , When however, it does not include within its scope and does not set itself to ask is whether that "how"'is "good or "bad" or whence that "how" ultimately derived.[17]

Making Value Judgements

Here Sherrington opens a fundamental topic: the role of *value judgments* in science. To claim that no value judgements are made in science would be ridiculous. Continuously the scientist selects one set of phenomena as the important one, prefers one, explanation over another, is pleased when a new "beatiful" theory appears to simplify what has previously been complex. Such judgements are based on simplicity, clarity economy and fruitfulness, that is, they are essentially aesthetic judgements. They clearly are not moral judgements like those involved in theology, ethics, and religions. As Sherrington observes later, since we assume the universe to be one vast harmonious unit, the scientist would be grossly presumptuous to apply to the universe value judgements which apply only to the behaviour of men toward men.

Teleology and Its Place in Science

Teleology, an approach or frame of mind in which we assert or desire that the world be shaped toward some purposeful end, is basically a philosophical position. Often in the sciences, especially biology, some teleological element is difficult to avoid. Cousider the grand implications of the theory of evolution, or the patterns of physical and chemical reactions. Many scientists, in a philosophical mood, will explore mentally the teleological implications of their work. As scientists, however, they avoid it. Telelogical arguments are not part of creative science. The reason is clear: If we spend our time contemplating the "ultimate design," we never get on which the business of creating increased predictability; we may even overlook implications of our observations which do not fit our teleological position, as so many of the very early scientists did. Teleological arguments do not lesd to observable predictions they often block science.

"The Problem" and "Problem Solving" as a Key Operation of the Scientist

Note how we heve apparently sidestepped problem solving as a 'key operation" of the scientist. At last, we must face it. Why do we place (almost at the end of our discussion) that which crops up first when people discuss the way of the scientist? The scientist is the problem solve *par excellence.* is he not? He begins his investigation with a problem, does he note? The problem comes first, does it not—even as teachers place it first on the board?

What we have tried to indicate is that problem solving begins even before the problem is stated; it begins perhaps unconsciously. The specific problem may be clarified when the scientist begins the empirical or the experimental investigative portion of his work; *but a great deal of his work goes on before he states his problem* in a formal way. The scientist solves problems; but this is not to say, we repeat, that he begins his investigation with a stated problem. If would seem, from our observations of the way scientists work that the problem filters our of a mixture of observations, flashes of insight ("eurekas"), vague, muddled dissatisfaction, tentative essay into empiricism, reading, consultation with others, thinking (conscious or unconscious), metal scanning and so forth. Eventually the problem does filter out, and an attempt to state it clearly is made. Clear statement is necessary if an experimental design or investigation is to be fruitful.

A scientist has long "office hours," in fact, they are almost continuous. Like all creative people, including teachers, he does not leave his problem in the locker at the end of the day. It is always with him, and clues or answers may appear at any time. This concentration or persistency is one hallmark of the embryo scientist. Science means hard work—as does all scholarship.

The kind of *problem solving* in which the scientist participates in order to develop concepts is certainly not to be equated with the *problem doing* found in workbooks. Problem doing is the fate of the classroom demonstration or experiment which must be done, finished, accomplished, and wrapped up in the 40 minutes of a class period. Hence our own predilection for distinguishing between the problem solving of the scientist (concept seeking) and the problem doing of the pupil (concept confirming). This is not to say that problem doing has no place in science teaching: indeed it has, especially when the history of science is taught (this is, actually the content of our present courses). But it is to say that the science class is also a place for problem solving or concept seeking. Concept attainment *can* be achieved in the science classroom.

It is, therefore, in order to emphasize the need to consider the nature of what we would have students do in problem solving, rather than prablem doing that we have placed this section in the middle of our discussion of the key operations of the scientist. We want, for our purposes, to distiuguish between the way of the scientist in problem solving (a creative act) and the routine of the usual problem doing of the student taking an established course, an which the solution is foreordained, because there is a rigid schedule (perhaps 40 minutes) and standard equipment. Do we want our students to engage in problem solving or in problem doing? We think both.

In problem *doing* the problem is stated first; it may even be handed down. In problem *solving* the statement of the problem is a creative act; it is a key operation resulting from much preliminary work in problem solving. It is the partial end result, as it were, of a part of the entire investigation. The scientist heaves as much of a sight of relief in clarifying his problem as he does in reaching a hypothesis, in developing a piece of crucial equipment, or in reaching his conclusion. The problem is a *result* of "sciencing," a result of the act of prolem solving, not necessarily the *cause* of it.

However, this is not to leave out the possibility of a problem occurring to a scientist in a flash of insight (the "eureka" process. Winning the Concept). But even then the problem will be refined, restated, and even discarded; as the scientist "does his damnedest with his mind, no holds barred," that is, as he develops a new relationship, as he attains the concept, as he solves his problem.

Recognizing the Limitations of Concept Attainment

Inevitably the scientist is haunted by his acute awareness of the limited information he has. He never knows "everything" about anything. Therefore his explanations are carefully phrased in terms of certain data when them selves are limited in precision. More accurate data man change the picture, as in the discovery of isotopes, and new lines of information from new tools may drastically modify the best answers created to date. The discovery of radio-activity certainly altered the earlier descriptions of atoms. Daltons's idea cf the atom as a solid particle underwent some remarkable transformations as Rutherford and Bohr began to "see" the atom as mostly empty space.

Scientific explanations should not involve assumptions contrary to basic conclusions widely accepted and used. For example, "flying saucers" were alleged to have been observed making abrupt 90-degree turns at speeds of 1,800 miles per hour. But such behaviour is inconsistent with fundamental, widely useful laws of angular momentum and inertia. The objects we know about just do not make abrupt 90-degree turns at such speeds. Until we do observe and confirm such phenomena many times, until we "know" such phenomena, we cannot accept them as part of our rational explanation of the universe. We would prefer to hold to the generalizations about angular momentum and inertia than to abandon them in favour of "little men from Mars," at least until we have more evidence. Our growing explanations of the world must not be inconsistent.

Concept Seeking—The Way of the Scientist and the Way of his World

Science, it seems, is more than empiricism, more than problem solving, certainly more than a method, or methods, even more than an attitude; it is a use of intelligence in a very complex, and at present little understood, cerebration in an attempt to make sense of this world. Its patterns of investigation and its operations in investigation are an attempt to discover the regularities, if any, of nature. The way of the scientist is designed to determine the way the world works. "Sciencing," as Bridgman has it, is a total operation. It really has no beginning as such and no end as such. It is, in a sense, the seeking of concepts—concepts and conceptual scheme which man build to help him understand man and the universe. One concept leads but to another.

In *On Understanding Science*,[13] Conant emphasized the cumulative nature of science in contrast to other fields of creative effort. Suppose Michelangelo. Raphael, Chardin, and Rembrandt were to come on the present scene to answer the question "Has painting advanced since our times?" We could visualize some interesting discussions, but hardly agreement. And what might Beethoven, Bach, Brahms, Schumann, and Wagner say of modern music?

What, however, would Newton, Galileo, Archimedas, Dalton, Vesalius, Grew, and Mendel say of modern science? Surely they would agree that science had advanced. And in analyzing why this advance might be so palpably clear, Conant derives what to him appears a consistently clear characteristic of scientific accomplishments: Science is both *cumulative* and *self-correcting*.

In examining the work of research scientists and in analyzing their reports. Conant observes that the *end result* of a scientist's work, if indeed, the word "result" may be used, is but another problem or several, not a conclusion or a "new" discovery, but a breathing space on the way to another concept. The scientist's way is an unending quest, unending conceptualization, or unending concept attainment; science is truly an "endless frontier."

If we wish to distidguish, in a broad sense, the research scientist from the applied scientist, let us say the doctor or engineer, then the notion of the unending quest becomes useful. The engineer has an end to `each specific task, for example, building a bridge; the doctor has an end in his "cured" patient; the technician has an end to his specific job with its completion.

The scientist's aim, conscious or otherwise, is a hunt for the conceptual scheme, for a spatial pattern in the infinite jigsaw puzzle of how the world works. A few such schemes (each based on many discrete facts, principles, and concepts) are given by way of illustration:

The earth is surrounded by an ocean of air.

Some diseases are caused by microorganisms.

Existing organisms are the result of evolutionary changes during the earth's history.

These conceptual schemes, which we admire and use resulted from observation and experiment, interwoven with creative mental effort. If observations and experiments are to lead to conceptual schemes which are to be a 'useful picture of the real world, they need to be reliable. But since man is not always reliable, the investigations must somehow be self-correcting. How is this self-correction by unremitting investigation built into the scientist's way of work?

All men are fallible, even scientists. They, however, are acutely aware of their failibility, as the quotation from Bridgman has shown, A major question then is how scientists dealing with incomplete and imperfect data are able to establish general statements on which they put great reliance. This is accomplished because scientists, conscious of their limitations, are inherently skeptical, in the best sense of the word, of their own work and that of others. Any single scientist's work must be confirmed. This open-endedness of an investigation, the realization that conclusions are not final, always provides opportunity for reconsideration of a result when new data become available. (New tools may play a major role in providing such new data). The scientist must accept some ambiguity in his knowledge even as he strives to lessen the ambiguity.

Bridgman has stressed the importance of using the correct operation in making observations. This involves the appropriate choice and arrangement of tools. We would all agree that a rather is appropriate for measuring height, but not for measuring intelligence. What constitutes the "correct" operation is always somewhat in doubt; one does the best he can and leaves to his colleagues now and letter the task of criticism and improvement.

Thus scientific work is never ended; it can always be extended and improved. For instance, Piltdown Man, long a debatable construct of the anthropologist, is now recognized as a classic hoax. How the fraud was finally exposed is a fine example of the self-correcting nature of scientific study. The original papers are worth reading.[19] They would serve well to illustrate many aspects of how the scientist works. They also illustrate how the scientist is constantly scrutinizing his "operations," "concepts," and "conceptual schemes." He is constantly asking:

What do we know?

How do we know it?.

How well do we know it?

As a scientist consistently asks questions of this nature, he introduces a self-correcting element into his ways of work. Scientists, we repeat, can and do make mistakes. Anyone who is familiar with the history of science can cite chapter and verse. But the scientist's way is self-correcting mainly because confirmation of observation is made by many others who are free, to a reasonable extent, of the personal bias which may have influenced the original statement. In science, too many cooks do not spoil the broth. The meal is prepared by many cooks working in many different kitchens. Hence any conclusion, confirmed as it is by different men, with different intent and in different situations, tends to approximate the "truth." And hence, when the conditions are better known and the operation is appropriate, future results are increasingly *predictable*,

Predictability is a characteristic sought not only by the scientist, but also by the cook who wishes to anticipate the hardness of a three-minute and of a four-minute egg. However, the scientist is not satisfied with merely knowing from past experience how hard the egg will be. He asks "how" this hardening occurs. From such questions, and extensive careful studies, have come conceptual schemes which are the basis of our knowledge of proteins, their properties and their molecular structure. Unlike many cooks, the scientist can explain what is happening and from that explanation can predict how other proteins will react. Furthermore, the scientist may be able to design and produce new proteins.

Attitudes of the Scientist

One student of ours thought he summed up the matter by saving. "I'ts the business of the scientist to be honest in his work." He might have added that be must also be patient enough to suspend judgement if only because it takes time to gather facts and to test their validity. It is unnecessary to assume that scientists are honest in all their affairs, or are open-minded (in the sense of suspended judgement) in all the relationships they must assume day in and day out. We are dealing with men, not supermen. We are dealing with the scientist *at work* as he searches for the discoverable regularities in nature.

Other fields of work require these same attitudes—the detective, the acountant, the judge *at work* need generous allotments of the "attitudes of the scientist." It is not these attitudes which make the scientist a scientist; it is attempting to work in science that forces these attitudes upon him.

Implicit in our whole discussion is the personal and social climate of opinion within which the individual scientist works. Certainly nowadays no scientist can work effectively shut off from contact with other scientists. What each one learns is important to others. Furthermore, informed colleagues serve as a sounding-board upon whom ideas are tested. Free exchange of information is essential. This has been called "academic freedom" but it is actually intellectual freedom for all. Also implied is the free "give and take" of criticism by a scientist's peers—others who are respected and trusted as equally well informed. The scientist is most effective in an intellectually free society.

In this atmosphere he is necessarily honest, courteous, generous, and objective. Wishful thinkers and charlatans are quickly recognized and discounted. Such a social climate, sometimes called the "climate of the laboratory," the "idea of a university," has evolved during some centuries because the criterion of accomplishment in science is objective, open for any and all to appraise. The ways of the scientist are the ways of intelligence.

Individual scientists, being human, are not perfect in all regards, even in their seientific work. History reports that even the wisest have been skeptical of ideas like evolution, relativity, and the quantum theory, which later proved to be of such major significance. But what is important to note is that such new, even radical, ideas have been open to further study. Not for many centuries has anyone attempted to prohibit further exploration. If free expression, free criticism, and objective evaluation of a man's efforts are the major ingredients which have prεmitted scientific work to be so successful over the past centuries, are they not the same ingredients tor which our whole society is searching?

Science as we shall use it in this book will be taken to mean, as Conant has developed it,[20] a series of conceptual schemes arising out of observation and experiment, and giving rise to further observation and experiment. And by *scientific methods* we shall mean all the operations, producers, devices, and types of processes by which scientists arrive at these conceptual schemes. These include observation, experiment, the "educated guess," the "chance discovery," library research, trial and error, common sense, verifying hypotheses, and many others.

Science is what scientists *do*, and most, if not all, do it because it is their way of living; it is fun. And there are likely to be as many operations, and hence voriations of methods and sequences of methods, as there are scientists. All of us may agree, however, that man is the only creature we know who appears able to use that "method of intelligence," the method which has resulted in the body of knowledge, of inventions, and of methods known as science.

Teaching the Ways of the Scientist

We began this chapter with a mention of 42 syllabuses at hand. Only one of them implies that there are many methods of science; 41 of them clearly imply that the scientific method is what we have called here "the empirical" or "try it and see" approach. Some of the syllabuses call this approach "problem solving." This tendency to simplify a complex subject, to seek a single approach to science, is understandable but it can be very misleading. Actually, as we have seen, *the highly empirical way of problem solving is but one approach of the scientist*, and it is a very slow one, used as a last resort when we have no better guides to action. And, of course, scientists seek and solve problems in order to find concepts, science might better be considered as concept seeking rather than problem solving.

The attitudes and methods of the scientist are *caught* as well as *taught*. How does one teach these things? How does the student learn to distinguish fact, hypothesis, theory, law? How does he learn *how* to define a problem, state his hypotheses clearly, and go on to design an experiment? Can the tactics and strategy of the scientist be taught at all? Chapter Next, and indeed the rest of this book, is given over to a discussion of these questions.

A Short Excursion into Developing One's Own Concept of Science and the Ways of the Scientist

1.1. As we noted at the beginning of this chapter, many qualified people have written volumes about the nature of science. Below we list a number of these books that you may find of interest. Perhaps from them, from our discussion about science, and from your own experience in working in science with children, you will want to construct your own definition of science. Perhaps you will be left as we have been, thinking that we we know what is scientifie and what is not, yet not being quite sure; for all the answers are not yet in. The edifice is not yet built: in our experience, edifices are subject to termites and shifting foundations—and a new architecture.

1.2. How might you create an interest in the air-pressure problem presented on pp. 15-16, in the following situations:

(a) Knowing that no student had ever seen it before?

(b) Knowing that every student had seen it before.

What possibilities does (a) present for prediction for introducing un-expected results, for starting the search for an explanation? What possibilities loes (b) present for testing a predictian, for exploring the significance of varying meteorological conditions (would the can collapse on the moon, on Many)? What other phenomena are consistent with, inconsistent with, the explanation?

1.3. Choose several other demonstrations and explain in a

similar way how they might be useful. (See the companion volumes.[21] which describe demonstrations, experiments, readings, field experiences which you may adapt to your own ways of teaching).

1.4. And written selections, like our selection from Redi, that might be used in class to develop an understanding of the ways of the seientist.

1.5. To what extent do you agree with this notion on the objectives a science teacher should follows?

"The teacher of science should aim at two main objectives:

(a) To get his pupils to reason about things they have observed, and to develop their powers of weighing and interpreting evidence.

(b) To acquaint his pupils with the broad lines of great scientific principipies, and with the ways in which these are exemplified in familar phenomena and applied in the service of man."[22]

1.6. Perhaps the following references will be useful to you in appraising your concept of science and the way of the scientist.

THE NATURE OF SCIENCE

Beveridgo, W.I.B., The Act of Scientific Investigation, rev. ed., N.Y.. W.W. Norton, 1957.

Boynton. I.L., The Beginnings of Modern Science, N.Y., Walter J. Black ("The Classics Clarify), 1948.

Bragg. W. Concerning the Nature of Things, N.Y., Dover Publications, 1954.

Bridgnate, P.W., Reflections of a Physicist, N.Y,, Philosophical Library, 1950.

Bronowsses J., The Common Sense of Science, Cambridge. Harvard U. Press, 1953.

Cannon. W.B., The Way of an Investigator, N.Y., W.W. Norton, 1945.

Cohen, L.B., Sience, Servant of Man, Boston. Little, Brown, 1948.

Conant, L.B., On Understanding Science, N.Y.. Mentor M-68, 1947.

——, Modern Science and Modern Man, N.Y.. Anchor A-10, 1952.

——, Science and Common Sense, New Haven: Yale U. Press, 1951.

——, Harvard Case Histories in Experimental Science, Nos. 1, 2, 6, 7, Cambridge: Harvard U. Press, 1950-53.

Enstein, Albert, and L. Infeld. The Evolution of Physics, N.Y., Simon & Schuster 1950.

Frank, P., Between Physics and Philosophy, Cambridge. Harvard U. Press, 1941.

——, Modern Science and Its Philosophy, Cambridge. Harvard U. Press, 1949.

——, The Philosophy of Science, Englewood Cliffs, N.J.: Prentice Hall, 1957.

Hall, E.W., Modern Science and Human Values, Princeton: Van Nostrand, 1956.

Holton, G., Introduction to Concept and Theories in Physical Science, Reading, Mass: Addison Wesley, 1953.

Nash, L., Atomic and Molecular Theory, Harvard Case No. 4, Comaridge: Harvard U. Press, 1950.

Roller, D.E., and D.H.D. Roller, Development of the Concept of the Electric Charge, Harvard Case No. 8, Cambridge: Harvard U. Press, 1954.

Sarton, G., History of Science and the New Humanism, Cambridge: Harvard U. Press, 1937, and N.Y.. George Braziller, Inc., 1956.

Seapley, H., et al., Readings in the Physical Sciences, Appleton-Century-Crofts. 1948.

Simpson, G.G., The Meaning of Evolution, N.Y.. Mentor M-66, 1949.

Whitehead A.N., Science and the Modern World, N.Y., Mentor M-28, 1925.

Wightman, W.P.D., Growth of Scientific Ideas, New Haven. Yale U. Press, 1951.

THE NATURE OF MATHEMATICS

Bell, E., Men of Mathematics, N.Y., Simon and Schuster, 1937.

Cajori, F., A History of Mathematics, N.Y.. Macmillan, 1924.

Courant, R., and H. Robbins, What Is Mathematics, N.Y., Oxford U, Press, 1943.

Dantzing. T., Number, the Language of Science, N.Y., Macmillan, 1949.

Davis, D.R., Teaching of Mathematics, Reading, Mass., Addison-Wesley, 1951.

Eves, H., An Introduction to the History of Mathematics, N.Y., Rinchart, 1953.

Freund, J.E., Modern Introduction to Mathematics, Englewood Cliffs, N.J., Prentice Hall, 1956.

Gamow, G., One, Two, Three, Infinity, N.Y., Mentor MD-97, 1947.

Kemeny, J.G., et al., Introduction to Finite Mathematics, Englewood Cliffs, N.J.. Prentice-Hall, 1957.

Kline, F., Elementary Mathematics from an Advanced Standpoint, N.Y., Macmillan, 1930.

Kline, M., Mathematics in Western Culture, N.Y., Oxford U. Press, 1953.

National Council of Teachers of Mathematics, Yearbooks, 1201 Sixteenth St., N.W., Washington 6, D.C.

Polya, Gyorgy, How to Solve It, N.Y., Anchor Books A-93, 1957.

——, Mathematics and Plausible Reasoning, Vols. 1 and 2. Princeton. Princeton U. Press, 1954.

Richardson, M., Fundamentals of Mathematics, N.Y., Macmillan, 1958.

Russell, Bertrand, Introduction to Mathematical Philosophy, N.Y., Macmillan, 1938.

Sanford, Vera, A Short History of Mathematics, Boston, Houghton Mifflin, 1930.

Sawyer, W.W., Mathematician's Delight, Baltimore, Pelican A-121, 1946.

——, Prelude to Mathematics, Baltimore. Pelican A-327, 1955.

Smith, D.E., History of Mathematics, Boston. Ginn. 1923.

Stabler, E., An Introduction to Mathematical Thought, Reading, Mass., Addison-Wesley, 1953.

Thompson, S.P., Calculus Made Easy, N.Y., Macmillan, 1914.

Wertheimer, Max, Productive Thinking, N.Y., Harper, 1945.

Whitchead, A.N., Introduction to Mathematics, N.Y., Henry Holt, Home University Library, No. 15, 1911.

THE HISTORY AND PHILOSOPHY OF SCIENCE

Barnett, L., The Universe and Dr. Einstein, N.Y., Mentor M-71, 1948.

Burtt, E.A., Metaphysical Foundations of Modern Science, N.Y., Anchor Books A-41, 1924, rev. 1932.

Butterfield, H., Origins of Modern Science, N.Y., Macmillan, 1951.

Campbell, N., What Is Science?, N.Y., Dover Publications, 1952.

Childe, G., Man Makes Himself, N.Y., Mentor M-64, 1936, 1944, 1951.

Childe, G., What Happened in History, Baltimore. Pelican p. 6, 1942, 1946.

Dampier W., A History of Science, Cambridge, Eng., Cambridge U. Press, and N.Y., Macmillan, 1949.

Dreyer, J. L., E., A History of Astronomy from Thales to Kepler, 2nd ed., rev. by W. H. Stahi, N.Y.. Dover Publications, 1952.

Frank, P., The Philosophy of Science, Englewood Cliffs, N.J., Prentice-Hall, 1957.

——, The Validation of Scientific Theories, Boston Press, 1954.

Hali, A.R., The Scientific Revolution, N.Y., Longmans, Green, 1954, and Boston, Beacon Press, 1956.

Jeans, J., The Growth of Physical Science, N.Y.. Cambridge U. Press, 1951.

Malinowski., R. Mugic, Science and Religion, N.Y., Anchor Books A-23, 1925.

Nordenskjold, E., The History of Biology, N.Y.. Tudor, 1931.

Oppenheimer, J.R., Science and the Common Understanding, N.Y., Simon & Schuster, 1953.

Partington, J., A Short History of Chemistry, N.Y.. St. Martin's Press, 1951.

Singer, C., A History of Biology, N.Y., Abelard-Schuman, 1950.

——, A History of Medicine, Oxford, Eng., Oxford U. Press, 1928.

——, A Short History of Science, Oxford, Eng., Oxford U. Press. 1911.

Sarton, G., A History of Science, Cambridge, Harvard U. Press, 1952.

Sedgewick, W., et al., A Short History of Science, N.Y., Macmillan, 1939.

Stimson, D., Scientists and Amateurs, A History of the Royal Society, N.Y.. Abelard-Schuman, 1948.

Taylor, F.S., A Short History of Science and Scientific Thought, N.Y., W.W. Norton, 1949, (Published as Science Past and Present, London, William Heinemann, 1945, in a short form).

Toulmin. S., The Philosophy of Science, An Introduction, London, Hutchinson's University Library, 1953.

Whitehead. A.N., Essays in Science and Philosophy, N.Y., Philosophical Library, 1947.

Wiener, P., Readings in Philosophy of Science, N.Y.. Scribner's 1953.

NOTES

1. P. W. Bridgman. "Prospect for Intelligence." *Yale Review*, *34*, 150, 1945, reprinted in *Reflection of a Physicist*, Philosophical Library, N.Y., 1950, p. 312.
2. P.W. Bridgman, on "Scientific Method." *The Teaching Scientist*, Dec. 1949, p. 23.
3. In this chapter we shall use the term 'concepts' in its dctisinary sense as equivalent to idea" on "principle" or "generalization." For its more technical mea ning.
4. We are the term concept attainment as it is used by Jerome S. Bruner, Jacqueline J. Goodnow and George S. Austin, in *A. Study of Thinking*, Wiley, N.Y., 1956. Essentially we are attempting to distinguish between the problem doing in which students in our established courses engage (which results in confirming concepts already discovered; these cannot be escaped if they follow instructions or the teacher's carefully directed questioning and the problem solving of scientists which is acutually directed at concept attainment. These concepts are new relationships. We discuss withim the limits of this work. the nature of concept seeking.

5. "Another look at the Ice Age'" *Science, 122*, 815, 1955.
6. James, B. Conant, *Science and Common Sense*, Yale U. Press, New Haven, 1951.
7. "Students in the high school science classroom are not, of course, consists" any more than one who writes composition is a writer" or one who draws students art class in an artist."
8. Taken from the board work of a teacher of science in "Hearsay" High School.
9. The accompanying volume by E. Morholt, P. Brandwein, and A. Joseph, *Teaching High School Science: A Sourcebook for the Biological Sciences*, details a number of such readiags. A good number of such cases are also given be:

   J.B., Conant, Duane Roller, Leonard Nash *et al., Harvard Case Studies*, Harvard A. Press, Cambridge, 1950-1953.
   I.B. Cohen *Science, Servant of Maths*. Little Brown, Beston, 1919.

   W.I.R. 'Beverdge', *The Art of Scientific Investigation*, W.W. Norton, N.Y., 1950.
10. G.G. Simpson. C.S, Pittendrigh, and L.H. Tiffany, *Life An Introduction to Biology*, Harcourt, Brace, N.Y., 1957, p. 22.
11. P.W. Bridgman *op. cit*.
12. For instance, Pasteur in his study of fermentation deliberately excluded the evidence of Liebig and Wohler that oil of bitter almonds was formed from amygdalim without the presence of living things. To Pasteur this occurrence was not "fermenation properly so called." We now know that Pastem was wrong but notc that without restricting the data to be considered he might never have achieved any solution. Later studies extended and thanged his results to include what had at first seemed hasically different.
13. John Dewey, *How We Think*. Heath. Boston, 1909: rev. 1933.
14. See, for example, W.B. Cannon. *The Way of an Investigator*, W.W. Norton. N.Y., 1945.
15. J.B. Conant, *On Understanding Science*, Yale U. Press New Haven. 1917.
16. Sir Charles Sherrington, *Man on His Nature*, Cambridge l Press N.Y. 1910 P. 2: we also our list of paperbacks. P. 531.
17. J.B. Conant, *op. cit*.
18. W.I., Straus, Jr., "The Great Piltdown Hoax," *Science*, 265, Feb. 26, 1954. See also K.P. Oaklev and J.S. Weiner, "Piltdown

Man," *American Scientist*, *42*, 573, Oct. 1955, and J.S. Weiner. *The Piltdown Forgery*, Oxford U. Press, N.Y., 1955.

19. J.B. Conant, *op. cit.*

20. Morbolt, Brandwein, and Joseph. Teaching High School Science. I Sourcebook for the Biological Sciences, Harcourt, Brace, N.Y., 1958, and Joseph, Brandwein, and Morbolt, Teaching High School Science. A Sourcebook for the Physical Sciences, Harcourt, Brace, N.Y. 1959.

21. John Brown, Teaching Science in Schools, U. of London Press, Ltd., London, 1943, p. 11, quoting I.J. Thompson.

2

Teaching the Ways of the Scientist

A note at the beginning: We may begin at the beginning by saying that the ways of the scientist, his method of intelligence, should be *caught*, not *taught*. Is this jargon, or does it have some meaning? It is a way of saying that attitudes and ways of thinking are sometimes most effectively taught by example and subtle practice: the learner "catches" them. These attitudes and ways of thinking are lived throughout the term of the relationship between teacher and student.

This doesn't mean that there is not to be a specified time, for example, a sequence of lessons, or even a unit, when the processes to be taught are planned consciously to be put emphatically before the students. But once that period of planned teaching is over, the processes are lived so that day in and day out, students and teacher practice them—that is, *do* them. In this sense, what is taught pervades the year's work.

More than a few experiences with honesty are necessary to motivate honest behaviour; an easily angered teacher cannot create calm, dispassionate discussion. Just so, a teacher who does not have the ways of the scientist embodied in his class work, as a part of his own way of sciencing, cannot develop effectively with a class the ways of the scientist. Here, as in so many situations, the old saying holds: "What you do speaks so loud I cannot hear what you say."

Thus we emphasize making the way of the scientist a continuous part of every course, every aspect of the curriculum. Later, in Section III, we shall study the curriculum in general science, biology, physics, chemistry, and physical science, as well as other possible courses. We shall discuss the special contribution of each course, its flavour as well as its structure.

Doing on Investigation in Class

If a biologist were turned loose in a chemist's laboratory he would not be ready to replace any of the chemists. But he would feel at home; surely he would not feel completely lost. This is to say that laboratories have an interior dooration, an odour, a design which strikes the same octave if not the same notes. Among the sciences there are certain ingredients common to all; these ingredients are the concomitants of the concept-seeking approach. For instance, the adherence to uncompromising accuracy, the careful statement of the purpose of the investigation, the preference for quantitative rather than qualitative predicsions and results, the mistrust of our senses and the course quant attempt to seek checks and balances, the search for the perfectly controlled condition in which only the postulated cause is being investigated, the self-correcting nature of the quest for verified hypotheses, and the attitude of the scientist himself. All these are the ingredients, or operations, of the scientific approach to investigation we seek to describe.

How may these ingredients of the scientist's way be practiced in the class-rooms. Here are some ways which have been observed to be successful. Naturally these are but examples. If we are correct in our position that fundamentals —whether they be skill in verbal expression, in the use of numerical expression, or in the method of intelligence—are taught pervasively throughout the school career, then whenever the opportunity arises, the method of intellegence will be demonstrated.

Teaching the "Appropriate" Operation

"Science" and "accuracy" have sometimes been called synonymous. This is so only if the appropriate operation is selected, and if quantitative methods are substituted for qualitative methods. (Bridgman, as has been mentioned before has stressed finding the "correct" operation as the essence of a scientific investigation). In demonstrating these ingredients of science, it is well not so obscues the point by making the situation too complex.

Some simple demonstrations to show the effect of the choice of the proper operation are:

1. Ask the students, "What is the temperature of the room?" Some will guess, others will suggest the use of a thermometer. Discuss why observing the thermometer is an appropriate operation if one wants a reliable figure.

2. Similarly, ask the students to determine the weight of an object by balancing it on the hand. List the weights which are the result of sensory impressions, then weigh the object. Discuss the importance of the appropriate operation.

3. Other occasions for discussing the importance of the appropriate operation occur:

   (a) When students first use a microscope, discuss the meaning of the word "invisible" with and without a microscope. Note the two possible meanings: "not yet seen" but possibly "see-able," in contrast to "not see-able" (x-rays?). With an electron microscope are virus practicles "seen",?

   (b) When students first use the telescape, or study its use, discuss the meaning of the word "visible" with and without a telescope.

(c) When students first use the Geiger counter, discuss the meaning of "seeing is believing" with and without a Geiger counted.

(d) When students first use indicators, discuss the possibility of measuring pH with and without indicators.

Teaching the Relevant Observation

Sometimes the way of the *scientist* is equated with the way of the *observer*. Yet it is a simple thing to note that the scientist's observations are not helterskelter or all-inclusive; they are selected and believed relevant; they serve the design of his experiment or they serve the purpose of solving his problem.

Students readily grasp the need for relevant observation. For instance, they would not expect the scientist who is studying polio to begin observing the structure of a bridge, or even the structure of an amoeba (unless he postulates that certain observable behaviour of the amoeba relates to the solution of his problem) as a part of this problem.

In one of our classes, students became interested in the problem "Is there any background radiation in the community?" In planning the method of investigation it was decided to *try* to find evidence of radiation in the community in which the students lived and *to see* whether, indeed, there was radiation. Some students suggested that the extent of radiation also be tested in a neighbouring community. Other students pointed out that, although this was interesting, it was not relevant to the solution of the immediate problem and that such observations would not help resolve the particular question to which an answer was being sought. There was agreement that this was indeed not relevant, but, since it was interesting, some students went on with it. Many similar illustrations of ways of teaching the notions of relevance in observation will occur to the teacher. Without appropriate operations, of course, there can be no relevant observations.

Teaching the Design of Pertinent Experiments

Many students of the scientist's ways are convinced that *the* operation which distinguishes him from any other type of investigator is an ability to design an experiment. Whichever way one looks at this ability, it seems to involve at least three initial operations:

1. Specifying clearly the purpose and plan of the study. This is within the area called "stating the problem," but, as we have sexn in Chapter 1, it is not quite that and yet more than that.
2. Suggesting one or more "possible conclusions," hypotheses which are tested. These are "educated guesses," the basis for "if . . ., then . . ." reasoning.
3. Designing an experiment, a planned series of observations under special conditions, in an attempt to test one or more of the hypotheses.

To illustrate, suppose a student were to say that exhaled air contains more carbon dioxide than does ordinary air. A profitable learning situation could be set up if the teacher were to ask whether anyone could design an experiment "to show that this is really so."

Were he to say "prove that this is so," as some teachers do, the student would resort to a textbook, or some other authority. This is a perfectly valid way of doing it. In fact most of the challenges hurled at students, such as "prove it," or "how do you know." are many times adequately answered "the books say so." To say "the book may be wrong" is hardly fair, because we do expect the student to use books as a basis for his work. And for most occasions in high school, carefully prepared textbooks are a useful base for science work.

Now we begin with specifying clearly what we want to determine. To do this, we may well ask students to state the "problem". In our experience, there is a useful way of teaching students how to state a problem clearly; we ask them *not only to state what is to be investigated, but also to state how it is to be done. By asking for the "operation," we ensure that the problem is clearly specified.* If only the "problem is stated," there is no assurance that what is to be investigated is more than a pattern of words. Assurance that the problem is understood comes when the student specifies what is to be done, that is, designs experiments.

Thus "Does exhaled air have more carbon dioxide than inhaled air?" is no assurance that the student knows what the words "exhaled air," "more than," "carbon dioxide," or "inhaled air" mean, even though he states a problem. But were he to add "Somehow we must devise an apparatus which will test in: certain specific manner the carbon dioxide content of the air we inhale and also of the air we exhale." he would be specifying clearly the purpose of the experiment and demonstrating his understanding of it.

How many different attributes of carbon dioxide might the teacher use as the basis for answering the question posed. If limewater turns milky, how does the student know this is evidence that carbon dioxide is present? In our experience, sketching the equipment needed for each possible test to be carried through for the qualitative-comparative result sought is a help in teaching students how to design the experiment.

In order to design an experiment, then, the experimenter generally goes through the process of doing the experiment in his imagination, postulates a likely cause, and then mentally tests the possible ways of determining whether this cause is actually the one responsible for the effect observed. In short, he solves the problem without "doing" the experiment; he "does it," as it were, in his brain. His conclusion is therefore hypotnetical; it is a possible conclusion, a trial explanation. In the exercise of hypothesizing, the student's knowledge, previous experience, imagination, and motivation come into play. Also acting is the teacher's skill in evoking interest Winning Participation). But in elaborating a "hypothesis" of a working idea, the student goes further than merely specifying the nature of the problem; he is "sciencing."

For instance, he may recall that cells use O_2 in oxidation and that CO_2 is the result of such oxidation; therefore, the expelled air must contain more CO_2 than the inhaled air. The postulated *cause* of the *effect* (increased CO_2 in exhaled air) is due to the activity of the cells. Therefore, if only he could obtain air which hasn't been placed in contact with living cells, and somehow test it *before* and *after*, he could compare inhaled and exhaled air for CO_2 content.

In designing the experiment, he bases the design on his "hypothesis." Somehow he must test air *before* it reaches the cells which change it and *after* it leaves the cells. In actual practice, if *sufficient time* is given, students will go through the stages and the reasoning mentioned above, and they will actually design an experimental device.

The student has in a sense designed an experiment with a *control.* A control, as students can learn to define it, is a comparative experiment with the *postulated cause removed.* The importance of teaching the *meaning* of the control cannot be overemphasized. For it is here that students sense the meaning of the *sine qua non* of science: *Scientists demand to know when they are wrong.* Testability of one's observations—hypotheses. conclusions, theories—therefore, is built into the design of an experiment.[1] Note that we use the term "design" to mean more than the selection and arrangement of hardware; design includes the chemicals and their possible reactions, the experimenter's activities during the run, selection of the particular phenomena to be noted, and the interpretation to be given if the results are as anticipated or to the contrary.

Teachning the Nuture of the Control

The control experiment is not necessarily a part of the design of every observation (it may be so well known that it need not repeated unless it becomes a very important part of the argument); but for clear knowledge scientists attempt to use the logic involved in a control experiment. Children may be taught the nature of the control quite early: elementary school pupils see its importance, and the general science course permits further development of its logical significance.

For instance, "Is moisture necessary to the growth of mold?" Youngsters will 'design an experiment in which they will expose bread and place it in a covered jar on a moist blotter. If they are then asked, "How do you know moisture is necessary?" they will eventually respond with a design in which the moisture is absent; that is, the *postulated cause* is removed from the control experiment.

Try them on similar hypothetical problems, for example:

A substance, called a catalyst (MnO_2—manganese dioxide), is said to speed up the production of oxygen from heated $KClO_3$ (potassium chlorate). Design an experiment to determine whether theis is so. Students will design an experiment in which some $KClO_3$ is heated gently with and without MnO_2. Howmuch MnO_2 gives the greatest yield of oxygen?

Many similar problems can be suggested (see the companion volumes in this series). Eventually students see that the difference between the *actual* experiment and *its control* is that in the control the *postulated cause* (moisture, or catalyst) has been removed, or held constant.

When students are given the opportunity to design experiments as a pervassive activity throughout the course, they soon begin to get the essence of the logic of the experiment.

Teaching the "Worthwhile" Experiment

In the introduction to this section, our two caricatures are drawn on the basis of what people think to be useful, meaningful, and worth while in teaching. One caricature of a teacher draws his problems strictly from *life*; the other draws his problems strictly from *subject matter*. What kind of experimentation is worth while—that drawn from problems deriving their essence from life, or from subject matter? We wish there were evidence, in the scientific sense, to support our viewpoint that teachers who draw from life the problems with which they deal in class do prepare students better for life than those who draw their problems from subject matter. But evidence in either direction is not available; the appropriate questions have not yet been asked experimentally

However, one may ask, and rightly so, whether a youngester's time can most profitably be spent in devising experimental designs to answer such question as:

What is the classification of the sea urchin?

What is the balanced equation for the preparation of Br_2?

What combination or pulleys will give us a mechanical advantage of 5?

or in answering such questions as:

What grade of engine oil is best used in winter or summer?

What is the relative effectiveness of different varieties of weed killers?

What is my blood type?

or in answering such questions as?

Can chrysanthemums flower in the spring?

Is fluoridation of water harmful?

Is heating by gas superior to heating by oil?

or in answering such questions as:

What kind of person shall I marry?

What policies shall we follow with regard to the use of atomic energy?

What shall be do with respect to race relations?

Is cancer curable?

What kinds of problems are chosen depends in part on the purposes of the school, its curriculum, and the training and philosophy of the teacher, among other considerations. But, in the main, we contend that the choice of an experimental investigation for the science class, as compared with the choice of investigation for the social science class, depends on what we consider to be the essence of science: namely, is the conclusion drawn from the experiment *testable*? As we considered in Chapter 1, the success of science has resulted from an insistence upon determining how well our answers described the world. Objective observational and experimental evidence allows all scientists to test each other's explanations. The more precise the predictions, the more precise is their testing and the clearer our acceptance or rejection of the explanations. Hunches and guesses about the working of the world are starting points, but the end is clear: We require evidence which will test our hunches. An "untestable" hypothesis is essentially meaningless, for we cannot know how much to rely upon it and under what conditions it may be useful.

In many other areas of human inquiry such clear tests are not often possible. In the prediction of the outcome of an election we never can be sure what would have happened if the candidates had done something differently. No experiment is possible, because each situation arises only once. The concepts of political science and many other areas are necessarily vaguar than those of chemistry or biology because the tests of their appropriateness are vague. This is the case not because those studying these difficult social affairs do not wish for clearer evidence and tests, but because the nature of the subject and of society is such that clear tests have not yet been devised. In these areas the "sense-making process" is more difficult, for controlled experimentation is usually impossible.

In science a critical question is then: *Are the predctioins testable*?

Teaching the Idea of "Numbers" of Experiments

The moment the teacher is convinced that students have an insight into the nature of the control, it is well to introduce the element of "numbers." This is statistical reliability, to be sure, but on a very simple level and scale. For instance, a biology teacher may state a problem somewhat like this: "A scientist wants to find out whether certain bacteria cause a disease. He has some guinea pigs which he knows are likely to get the disease. What should he do?"

The students will generally say. "Let him inject one guinea pig with the bacteria, and not inject another, the control." (Later, students may learn such refinements as the use of litter mates, i.e., genetically constant material.) So far so good; they "know" the nature of the control.

Now the teacher asks, "Can you be sure that all other guinea pigs will react in the same way as this one does?" Students will generally suggest another pair of guinea pigs. In a little while, students realize the need for a sufficient number of experiments in order for an experimenter to come to dependable conclusions. When we deal with laboratory work, we shall see that one of the advantages of boratory work is that the entire class confirms its own experiments by designing them with adequate controls and in adequate number.

However, visualize almost any classroom in this country, or in any other, if the accounts of observers who travel internationally are correct; we are likely to see something like this:

Chlorophyll is removed from one geranium leaf. When the leaf is stained with iodine, it turns black—a test for starch. Conclusion to be accepted by the students: *all* green plants make starch.

But we have only one test of one leaf; we do not know anything about *all* green leaves. Only after testing many kinds of leaves—red oak, pine needless, elodea, algae, moss, and so on—might we hazard a guess that "probably" all green leaves make starch. A laboratory exists for such explorations.

The position of a teacher who tells students. "You can accept the generalization on my say-so," or "the textbook's say-so," is awkward indeed. This is a kind of authoritarianism, yet the scientist remains the enemy of all authoritarianism.

If the reader thinks back to his experience in the high school classroom. he will probably recall that demonstrations (they were called "experiments") were performed *once*. Wasn't a conclusion generally reached from *one* demonstration? Isn't this commonly true of science teaching today?

It is, unfortunately, a common occurrence. Yet most science teachers believe and say that, in tne establishment of scientific information, the phenomenon should be observed many times under similar, and then under different, deliberately changed, conditions. As Brownell and Hendrickson[2] have said." "No child should be permitted from a single laboratory observation to formulate a generalization. . . . Instead, both in deriving a generalization and in applying it, learners need to encounter other instances of its validity: and the more practical, concrete, and natural or everyday such encounters are, the better." They also said:[3] "Children need to know not only what concepts *are* (what they include, to what they refer) but also what they are *not*. The time it takes to clear up misunderstandings is amply repaid."

If conclusions are continually elicited solely on the basis of one observation, one trial, we cannot expect the pupils to see the need for, or believe, that scientists actually conduct many experiments, make many observations, and are cautions in coming to conclusions. The difference between "what I do" and "what I say" does not sit well when students must come to accept (believe) a fair-sized conclusion every hour.

Yet what is the teacher to do when he is always working within a limited time? How can he immerse the class in *one* major activity of scientists—that of designing an experiment, then checking predictions by doing it? How can he rid his instruction of chalk-talk and avoid the error of the single shot activity leading to an all-embracing conclusion?

A tentative solution—use your laboratory. The first few classroom demonstrations in the year can be made the occasion for developing the idea that each represents many similar ones that have been done before. At that time the point can be made that especially important problems will be studied more fully in the laboratory, with possible variations in the kind or in the amount of materials. This statement must, of course, be followed up with the performance it promises. The teacher can stress that a demonstration followed by individual laboratory work or not, is only an *example* of how scientists work. Time and materials permitting, the teacher should attempt to sumbit the most important problems to fuller laboratory experimentation.

As students continue to design experiments throughout their high school experience in science, they begin to get a glimmer into the ways of the scientist. They get a glimpse of the scientist at work, or our "method of intelligence."

Time to Investigate—Time to Design and do Experiments

Any science teacher who daily meets a class and yearly confronts an overcrowded syllabus will wonder when student may have the time to design and do the kind of experiment we have been talking about. There are several answers relating to practicable modificaiions of present courses of study to give more time to leisurely teaching—a prerequisite and perquisite of effective science teaching. The modifications will be dealt with in various sections in this book. Nevertheless, the teacher who is aware of the opportunities in class situations may grasp such opportunities as often as he wishes. The following is such an instance.

The "Problem" Situation; the Opportunity Grasped

Sometimes, and it can happens at least once during the

year's work, a problem situation occurs. Then a teacher who believes in teaching "science" as well as "history of science" grasps the opportunity. So do his youngesters. Here is the story of such an occasion, described in an "open letter" received by one of the authors.

The June, 1948, preliminary Science Regents contained the following question:

"Explain how you could test the truth of one of the following statements:

(a) Cut flowers will not fade so rapidly if an aspirin tablet is discovered in the water in which they are kept.

(b) How water freezes sooner than cold water.

(c) Water heats faster in a black kettie than it does in an aluminum kettle."

As a follow-up, the students in my 8th grade classes set up experiments at home to test part (b) 81 per cent of the students reported that for complete freezing but water froze first. Out of curiosity I continued the experiment with succeeiding classes. Each time I increased the controls and eliminated possible errors. The results were always the same. From 85 to 95 per cent of the pupils found that the hot water froze sooner.

The experiment was performed in two ways:

1. Equal volumes of boiling (100°C) and cold (18°C) water were placed in separate ice trays, set side by side, and tested. The experiment was repeated with the positions of the hot and cold water trays reversed.
2. Equal volumes of water were tested as above first cold, then hot.

The evidence for the hot water freezing sooner is as follows:

1. 903 students performed the experiment. They were instructed in revealing the meniscus of a liquid, in checking temperatures of both liquids and the interview of the freezing compartment, in controlling the checking procedures, in using different amounts of water, etc. 872 students found that the hot water froze carlier.

2. With the aid of the home economics department, experiments under laboratory conditions were performed. Here are some of the typical results:

Volume (ml)	*Temp. (°C)*	*Weight (gm)*	*Time (min)*	*Freezing temp. (°C)*	*Ice weight (gm)*	*Remelted volume (ml)*
50	72	18.20	09.15	1.8	17.5	47
50	19	19.45	10.01	0.3	19.1	49
100	70	96.35	17.50	1.7	94.2	95
100	19	99.20	21.04	0.2	97.3	96

3. I checked through scores of chemistry and physics texts. There was no mention of this particular problem, although one book. *I Brief Course in Physics* by Earls (Prentice-Hall, 1949), stated that the freezing point of boiled water was higher than that of cold water which, of course, contained dissolved air.

4. Hot water weighs less than cold water. There are less grams to freeze.

5. Rapid evaporation, hence rapid cooling, proceeds quite vigorously above the hot water ice trays because it presents a large surface area and because the water is quite hot. This leads to two assumptions:

   (a) In the early stages of cooling the hot water is aided in its "chase" of the "handicapped" cold water by rapid evaporation.

(b) Some of the hot water is lost to the sides of the freezer compartment in the form of frost, thus further reducing the number of grams in the hot water tray.

6. Tables in the *Handbooks of Chemistry and Physics*, 1954-55, page 20-24 show that the thermal capacity for air free water is not the some at each degree:

Temp. (°C)	*Thermal capacity (cal/gm/C°)*
0	1.00738
5	1.00368
55	.99895
64	.99988
100	1.00697

It would appear that the hot water would be given a "cooling boost" as it fell in temperature through the middle degrees.

In any event, the class concluded that, under similar conditions, equal amounts of hot water freeze sooner than equal quantities of cold water.

Kenneth Dunn, Science Department
Carle Place High School, Carle Place,
New York

(Perhaps the following comment may be relevant: Hot water will melt through the film of ice on the freezing tray and will make better contact with the freezing unit. The cold water tray may be insulated by a layer of ice which has a lower thermal conductivity than does the material of the freezing shelf.)

Kenneth Dunn grasped the opportunity to teach the design of an experiment. The class, as we understand it, worked hard but the students also had fun. Science is fun, as well as hard work; it is also an important human activity.

Observing a Total Scientific Operation

Total scientific operations can be observed only where they are in process; and they are going on all the time in the laboratories of the nation. It may be difficult for most youngsters to observe a scientist at work, yet it can be done.

Observing Scientists "in the Flesh"

Wherever students get the opportunity to meet scientists or to work for them, the experience is of value to both. The scientists involved learn about the schools and are stimulated to help the science teacher by offertng equipment and by appearing before science clubs and student assemblies. In one school situated near a university, students acted as laboratory assistants in minor tasks at first, and later, in advanced ones. In another school situated near a hospital, students assisted part-time in the laboratories. Simslarly, in several schools located in large cities where national or state conventions of scientists were held, the students were regularly sent to the meetings. In all these cases they reported on the activities of the scienties for their classmates, science club members, the school paper, and the school magazines. Also, one of the scientists at such a convention, or from such a laboratory, may be willing to come to the school and speak at an assembly.

One of the most fruitful ways for students to get direct information about the activities of scientisis is to ask a former student who is majoring in science at a university to return from time to time and report on the activities there. Former students in science returning to their previous haunts have a healthy and stimulating effect on the student in school, for developing strong motivation is still a function of the school.

Observing Scientists "Vicariously"

As has been stated, it is not always possible to study the scientist and his ways directly, but it is usually possible to study his ways by studying his life through a biography, a film, or the "case study" approach. By these means it is possible to study the way scientists gets problems, fashion their grand and limited working hypotheses, design their experiments and develop their conceptual schemes, and evolve the questions they ask of nature. After all, the most important attribute of the scientist is not that he finds the right answers but that he finds the right questions. (As we developed in Chapter 1, this activity of finding the right questions occurs in the brain; and there is no present verifiable knowledge as to how it occurs.)

Through Biographies. Reading biographies of scientists is apparently very useful in getting at their way of approaching and solving problems. For instance. Hans Zinsser's account of his life in *As I Remember Him*, Vallery-Radot's *The Life of Louis Pasteur*, or Eve Curie's account of her mother's life, *Madame Curie*, help the student glimpse the tragedies and triumphs of a scientist's life. They get a bit of the smell of the laboratory, they see imagination and gather information, they get an inkling of the scientist as scientist and as human being.

Through motion pictures. It is now possible to get several films useful in grying youngsters a miniature of the scientist. For instance, *Louis Pasteur*, *The Magic Bullet* (a vignette of Paul Ehrlich), or *History of Chemistry* will help youngsters to get some insight into the nature of scientific work.

Lists of such films may be found in the companion volumes (see footnote ca. p. 33); some of these films are available on loan.

By a study of their papers. One of the methods the scientist uses, highly typical of his approach, is publication of data. He intends, unless he is bound by security regulations, to make his knowledge available to all. And he describes his methods in such a way (in a "formal logic") that they may be confirmed by others. Hence his paper is part of the self-correcting method of the scientist.

It is entirely possible (and it has been done) for a teacher to get a set of research papers by writing to scientists asking for reprints. However, he should make his requests reasonable, and remember that someone, often the scientist, had to pay for the reprints. (Perhaps he need request only one, and can mimeograph copies.) These reprints may be examined in class and the organization discussed. For instance, general headings such as:

1. Introduction,
2. Method,
3. Observation,
4. Conclusions, and
5. Bibliography.

As a result of a study of such papers youngsters "catch" the climate of accuracy, carefully detailed work, and essential honesty of the scientist's efforts. They catch, in short, something of the *attitude* of the seientist. They can see something of the way the scientist uses guesswork, common sense, trial and error, as well as "imagination" to get at a solution to the problem. They will also see, if the reprints are wisely chosen, that unsuccessful as well as successful experiments are reported.

Through the case method. Students of law or business administration often study cases dealing with specific important generalizations and analytical operations in these fields. Students of science can also profit from the use of case studies in science. In fact, several large universities, including Chicago, Colgate, and Harvard, have used cases for some of their general education courses in science. The intent is to have the student, through the study of relatively simple factual material, follow closely from the original writings the way in which the scientist tried to reshape his material and ideas to reach some major conclusion. Conant has called this "looking over the shoulder of the scientist at work." The student should recognize through his own experience the difficulties and procedures of the scientist at work.

Several high school science teachers have also for many years used cases in their instruction. The materials chosen are less lengthy and complex than those used in college, but the intent is the same.[4] A series of experimental cases for use in high schools is now being developed. The first titles are: *The Sexuality of Plants, The Discovery of Bromine*, and *How Far Away is the Sun*? Each is chosen to illustrate clearly certain aspects of scientific creativity. A number of high school texts also include sample case histories.

The Harvard Case Histories, available in pamphlet form,[5] are:

1. J.B. Conant, *Robert Boyle's Experiments in Pneumatics*
2. J.B. Conant, *The Overthrow of the Phiogiston Theory*
3. Daune Roller, *The Early Development of the Concepts of Temperature and Heat*
4. L.K. Nash, *The Atomic-Molecular Theory*
5. L.K. Nash, *Plants and the Atmosphere*
6. J.B. Conant, *Pasteur's Study of Fermentation*
7. J.B. Conant, *Pastenr's and Tyndall's Study of Spontaneous Generation*
8. Duane Roller and D.H.D. Roller. *The Development of the Concept of Electric Charge.*

While these cases are likely to be too complex for secondary school pupils, they are admirable for use in teacher training institutions.[6]

Each case study is built around significant sections from the original writings of the scientists involved. There is an analysis of the historical background behind each investigation and a painstaking account and analysis of the investigator's purpose, his way of recognizing the problem, his grand and limited working hypotheses, his experimental design, his techniques, his errors, and his difficulties.

The essential way of dealing with the case study is to have students read it before class, then "milk it dry" with a discussion of the ways of the specific scientist under scrutiny. At Forest Hills High School, for example, George Schwartz uses mimeographed excerpts of the actual papers of the scientists involved in the development of the concept being discussed. For instance, in a discussion of spontaneous generation, his students are given a translation of a portion of Francesco Redi's experiment and a part of Pasteur's as follows:

PASTEUR PROVES THAT FERMEHTATION AND PUTREFACTION ARE BROUGHT ABOUT NOT BY THE AIR BUT BY PARTICLES SUSPENDED THEREIN[7]

I believe I have rigorously established in the preceeding chapters that the organized productions of infusions that have previously been heated have no other origin that the solid particles which the air always carries and coastantly less fall on all objects.

If there can still remain in the mind of the reader even the least doubt in this regard, it will be removed by the experiments of which I am about to speak.

I place in a round glass flask one of the following liquids, all of which are very easily altered by contact with ordinary air: watery extract of yeast, sweetened watery extract of yeast, urine, beetroot juice, watery extract of pepper; then I draw out, at the blowpipe, the neck of the flask so as to give it various curvatures as is indicated [accompanying figures]. I then bring the liquid to a boil for several minutes until the stream issues freely from the open end of the drawn-out neck, using no other precaution. I then let the flask cool. It is a remarkable fact, calculated to astonish anybody who is used to the delicacy of experiments relative to so-called spontaneous generations, that the liquid in the flask remains indefinitely without alteration. One can handle the flask without apprehension, move it from one place to another, allow it to undergo all the temperature changes of the seasons, and the liquid in it shows not the slightest alteration and keeps its smell and taste. . . . There will be no change in its nature other than, in certain cases, a direct oxidation, purely chemical, of the matter. But we have seen by the analyses which I have published in this memoir how this action of oxygen is limited, and that in no case are any organized productions developed in the liquids.

It appears that the ordinary air, returning with force at the first moment, must reach the tlask quite unaltered. That is true, but it encounters a liquid that is still near its boiling point. The re-entry of the air thereafter occurs more slowly, and by the time that the liquid is so far cooled as no longer to be able to deprive germs of their vitality, the re-entry of the air becomes so slow that it leaves behind, in the damp curves of the neck, all the dust particles capable of acting on the infusions and bringing about the production of organism. At least, I see no other possible explanation of these curious experiments. If, after one or many months' stay in the incubator, the neck of the flask be detached by a nick with a file. without touching the flask in any other way . . . after twenty-four, thirty-six or forty-eight hours, molds and infusoria begin to show themselves absolutely in the ordinary way or just as if one had sown dust particles from the air in the flask.

In class, a discussion brings out the essential purpuse, method, and out come of the experiments, as well as the way of the scientist—all this, of course, in addition to background reading in the text.

For instance, note how the spirit of Jan Ingen-Housz's investigations is disclosed even on a brief reading of the excepts which follow:

EXPERIMEMTS UPON VEGETABLES DISCOVERING THEIR GREAT POWER OF PURIFYING THE COMMON AIR IN THE SUNSHINE, AND OF INJURICOTT IT IN THE SHADE AND AT NIGHT[9]

I was not long engaged in enquiry before I saw a most

important scene opened to my view: I observed that plants not only have a faculty to correct bad air in six or ten days by growing in it, as the experiments of Dr. Priestley indicate, but that they perform this important office in a complete manner in a few hours: that this wounderful operation is by no means owing to the vegetation of the plant, but to the influence of the light of the sun upon the plant . . . that this operation is far from being carried on constantly, but begins only after the sun has for some time made his appearance above the horizon, and has, by his influence, prepared the plants to begin anew their beneficial operation upon the air, and thus upon the animal operation, which was stopt during the darkness of night: that this operation of the plants is more or less brick in proportion to the clearness of the day, and the exposition of the plants more or less adopted to receive the direct influence of that great luminary; that plants shaded by high buildings, or growing under a dark shade of other plants, do not perform this office, but, on the contrary, throw out an air hurtful to animals, and even contaminate the air which surrounds them; that this operation of plants diminishes toward the close of the day and ceases entirely at sunset, except in a few plants, which continue this duty somewhat longer than others; that this office is not performed by the whole plant, but only by the leaves and the green stalks that support them; . . . .

This work is a part of the result of above 500 experiments, all of which were made in less than three months. having begun them in June and finished them in the beginning of September, working from morning till night.

EXPERIMENTS SHOWING THAT PLANTS HAVE A REMARKABLE POWER TO CORRECT BAD AIR IN THE DAY

56. A sprig of peppermint put in a jar full of air fouled by breathing (so as to extinguish a candle), and exposed to the sun, had corrected this air in three hours so far that a candle could burn in it.

57. A sprig of nettle was put in a jar full of air fouled by

breathing so as to extinguish a candle; it was placed in a room during the whole night: next morning the air was found as bad as before. The jar was put at 9 in the morning in the sunshine; in the space of two hours the air was so much corrected that it was found to be nearly as good as common air.

EXPERIMENTS SHOWING THAT THE SUN, BY ITSELF WITHOUT THE ASSISTENCE OF PLANTS, DOES NOT IMPROVE AIR BUT RENDERS IT RATHER WORSE

12 Two jars, half full of air taken from the atmosphere at the same time, and half full of pump water, were left by themselves during four hours: the one was exposed to a bright sunshine, the other placed within the house, only two steps from a door opening in the garden. The air kept in the house gave, in six diffirent trials, constantly the appearance of being better than that of the jar placed in the sun. One measure of the air kept within doors with one of nitrous air occupied 1.06½ per cent, whereas that exposed to the sun occupied 1.02½. I must, however, acknowledge that this experiment ought to be repeated more than once, to put the fact out of any doubt. I made it the very last day of my stay in the country, and thus had no time to repeat it.

Experience with a number of teachers has indicated that the case study method is exceedingly useful especially when library facilities are available. Further discussion of methods of teaching science in later chapters will offer further approaches which make use of the case method.

Through scientists' evaluations of their own methods. Scientists are not illiterate. In fact, many times they make statements which describe how they think, feel, appraise, or evaluate their own methods of work. Sometimes these evaluations are brief but comprehensive statements. An examination of even a few brief quotations will indicate that valuable discussions may result from their analyses.

WHAT IS SCIENCE?
ITS SOURCES, AIMS, AND PROCEDURES[9]

This is the first fact about science: it is a human activity. As such it is subject to all the frailties of human endeavour. Its history is a history of errors, slowly corrected a history of effort, a search for a wider knowledge of facts and deeper, more embracing schemes for understanding and explaining them. Science is not a cold-infalliable, impersonal machine; it is not just a collection of formulas and technical terms about abstruse questions remotely related to everyday life and thought. Science is ultimately refirmed common sense. It is the organized knowledge developed and possessed by men and women of the world they live in.

As a human endeavour, science is dependent on human motivation. The motivations for scientific endeavour are not hard to find. First, there is the ever-present need for knowledge that helps man master the environment he lives in, making it subservient to his needs. Our knowledge of astronomy began its growth impelled by the need of a calendar for the season-conscious Babylonian farmer. In our own day, the growth of knowledge of nuclear physics has been impelled in part by unhappier but no less practical considerations.

Secondly, and in the long run more important, there is the almost universal curiosity in men which seeks satisfaction in knowledge of the world for its own sake knowledge not only of the facts themselves but of how they are related and how things are caused; in short, knowledge of how the world of fact can be explained and so understood. The simple and naive questions of the small child when they recur in mature minds are the life's blood of science. "Who dug it?" as a question on first seeing the Grand Canyon is naive only in phraseology; in the older man this child-question leads to our understanding of the catting power of running water. Cats may be killed by curiosity, but science dies for the lack of it.

Or a view of mathematics:[10]

> Mathematics, rightly viewed, possesses not only truth, but supreme beauty—a beauty cold and austere, like that of sculpture, without appeal to any part of our weaker nature, without the gorgeous trappings of paintings or music, yet sublimely pure, and capable of a stern perfection such as only the greatest art can show. The true spirit of delight, the exaltation, the sense of being more than Man, which is the touchstone of the highest excellence, is to be found in mathematics as surely as in poetry. What is best in mathematics deserves not merely to be learnt as a task, but to be assimilated as a part of daily thought, and brought again and again before the mind with ever-renewed encouragement. Real life is, to most men, a long second best, a perpetual compromise between the ideal and the possible; but the world of pure reason knows no compromise, no practical limitations, no barrier to the creative activity embodying in splendid edifices the passionate aspiration after the perfect from which all great work springs.

Or even a very brief remark, such as the following:[11]

> The white man drew a small circle in the sand and told the red man, "This is what the Indian knows," and drawing a big circle around the small one, "This is what the white man knows." The Indian took the stick and swept an immense ring around both circles: "This is where the white man and the red man know nothing."

The student as "scientist"

High school students are not scientists, but they are scientists in the making. Some of them will be scientists; all of them as youngsters or adults will use some of the ways of the scientist.

If the ways of the scientist are taught pervasively during the year, the student will have the opportunity to do the various things which characterize the scientist. He will be given the opportunity to:

1. Use his imagination to develop ideas (even wild ones; wild ideas eventually succumb to the quiet of thought).

2. Design experiments, and in so doing learn to

   (a) specify his problems,
   (b) clarify his hypotheses,
   (c) rectify his observations,
   (d) verify his conclusions.

3. Give reports to his classmates (a seminar situation).

4. Do "research" in the library and the laboratory.

5. Write a pape.

6. Report before his peers (in the science class or in a Science Congress).

7. Exhibit before his peers (in a Science Fair).

The teaching method and techniques by which these are to be accomplished belong properly in Section III of this book. A course in advanced science wherein students do "research" parojects will be described.

Here it is sufficient to indicate that a good method, possibly the best, of helping the student to understand the way of the scientist is to have him live the life of the scientist, as best he can.

Scientists and Science Teachers as Human beings

As has been said, scientists have been successful. Although this success has been tinged with remorse for some of the social uses to which scientific discoveries have been put, still it is envied. So we find for instance, advertising becoming "scientific"; toothpastes, lipsticks, cigarettes must somehow be associated with *Science*.

The result of all this has been to forget that the scientist is a human being as well, happy and troubled, with the goals and aspirations of all human beings. Indeed Kubie[12] has emphasized the price a scientist might pay for his choice of a life of research.

> Scientific research is conducted largely behind closed doors, and the accuracy of any man's observations and the veracity of his reports depend ultimately upon his honesty. This honesty depends in turn upon maturity, upon some degree of security, and upon a sense of identification and fellowship with competitors. Under present conditions, it is a tribute to scientists that violations of their code of honour are so rare that when lapses occur they become historic scandals. This issue is especially delicate in such fields of science as psychology, psychiatry, and psychoanalysis in which it is difficult to repeat another man's observations for purposes of objective clinical or experimental or statistical confirmation.
>
> . . . Certainly the idyllic picture of the innocent, child-like scientist who lives a life of simple, secure, peaceful, dignified contemplation has become an unreal fantasy. Instead, the emotional stresses of his career have increased to a point where only men of exceptional emotional maturity and stability can stand up to them for long, and remain clear-headed and generous-hearted under such psychologically unhygienic conditions. Thoughtful educators are beginning to realize that the socio-economic basis of the life of the scientist must be entirely overhauled; that the psychological setting of his life needs drastic revision; and that, at the same time, the emotional preparation for a life of research is at least as important as is the intellectual training.

In any event, for the teacher this means that the must assume the responsibility for guidance so that youngsters might know the kind of life the scientist leads. Biographies sometimes hide the daily routine the tediousness of investigation; they emphasize results rather than the thousands of disappointing failures and frustrations. Invite an investigator to talk condidly to students—to those who want to be scientists, and to those who do not. The latter need to understand the kind of man who gives his life over to investigation.

As we noted earlies, perhaps students who are now in graduate school and who have returned for a visit will speak to classes in science. In this way, a line of communication can be maintained between youngsters preparing to go into science, those who will need to understand the nature of science and scientists in a "Scientific" age, and those who are already in the thick of investigation.

A word about transfer of training:. There is little evidence that the need and the act of becoming a scientist per se makes a youngster a better person. Indeed, it seems that youngsters in high school who intend to become scientists have already achieved a type of "behaviour pattern" which predisposes them to the intellectual wave lengths of life.[13] It appears, too, that youngsters who enter upon an intellectual life must develop bread-and-butter behaviours and attitudes which are part of the pattern of life they have assumed; hence, they are bound eventually to differ from those who have, let us say, assumed the pattern necessary to success in business.[14]

Nevertheless, it is conceivably sound to say that the science teacher plays a part in helping the youngster identify himself with a pattern of behaviour which has been identified with the scientist.

The teacher, being *in loco parentis*, can at least approach the pattern of living he seeks to establish in the lives of youngsters. There are indeed attitudes which are helpful for scientists to assume, if only to help them get at their investigations with a minimum of emotional friction. These attitudes have been generally grouped together as "scientific attitudes." Actually they are characteristic of the judicial temperament. They are part of the method of intelligence we have been talkihg about. Surely it is intelligent to suspend judgement, to be open-minded, to be honest and patient when one attempts to solve any problems.

It is specious, then, to allot these attributes to the scientist alone when even casual thought will show them to be necessary to wise, gracious, and balanced living. Nevertheless, teacher who is not open-minded, who is bigoted and prejudiced, who is not honest, judicious, and patient in his dealings with his students and the problems in class, cannot with propriety "teach" these attitudes or help make them active attributes. To make a mistake is human, but to admit it is also human, and some would say it is the better part of making the mistake. The fact is that scientists do make mistakes, as scientists and as human beings. As human beings their method of intelligence is fast becoming a goal and aspiration of right action. As scientists their mistakes are corrected, for their method of intelligence serves them in an undertaking where the free even joyful, experience in search of meaning is the central goal, and where the atmosphere of self-correction prevails. They breathe, in short, the climate of science.

This atmosphere, this climate, is also that of the science classroom. Where the method of intelligence prevails, science is "caught" as well as "taught." The science teacher clearly has prime responsibility for what goes on in his classroom; hence he has responsibility for the climate in which he teaches.

A Short Excursion into Appraising Activities Appropriate to Developing a Climate of Science

Four laboratory exercises and experiments are offered below for your evaluation.

(a) *Measuring Buoyancy*

1. Set up the apparatus as shown on the blackboard. Use any weight such as a rock or a metal object that fits into a can. Measure the weight of the object. Its weight is........grams. Fill the can with water until it pours over the spout, Pour out the excess that flows

into the overflow can. Lower the object into the water as shown in B and note its apprent weight. The apparent weight in water is .... The object seems to have lost .. .. ..of weight. The buoyancy of the water on the rock is .. .. What seems to be the connection between buoyancy and loss of weight?......

2. Weigh the water displaced by the object as follows: The weight of the each bucket with the overflow water displaced by the object is......... The weight of the catch bucket when empty is .. .. How does this weight of overflow (displaced) water compare with the buoyancy on the object? ....

3. Repeat the experiment for another object. The weight of the new object is......... The apparent weight in water is .... The loss of weight (buoyancy) is .... The weight of the catch bucker with displaced water is......... The weight of the catch bucket alone is........ Therefore the weight of the water displaced by the object is .... How do the buoyancy and weight of displaced water compare?........

4. Repeat the above experiment, using the same object as in the previous experiment but with some other liquid such as salt solution or alcohol. The weight of the object is .... The apparent weight of the object in the liquid is .... The buoyancy (loss of weight) is .... The weight of catch bucket and displaced iquid is ....The catch bucket alone weighs........ Therefore, the displaced liquid weighs....... . How does the weight of the displaced liquid compare with the buoyancy of the liquid on the object? ....

   Arriving at your conclusions. State a hypothesis connecting the buoyancy of a liquid on an object immersed in it and the weight of liquid displaced by the object ....

(b) *What is a Control Experiment*?

Is moisture needed for mold to grow on bread? Set up

an experiment with a *control* to answer the question above.

Here are some suggestions. My procedure:

1. Expose a slice of bread to the air until the bread is dry.
2. Break the bread into two pieces.
3. Moisten one piece; leave one piece dry.
4. ....
5. ....

My observations of the results:..

My conclusions:..

(c) *What Conditions are Favourable for the Spoilage of milk*?

The plan for my experiment...

....

....

What happened:...

....

Conclusion:*..

....

*Did classmates with experimental plans like yours get the same results? It is good procedure to check results with the conclusions of others. Scientists do not accept conclusions which come from doing an experiment just once.

(d) The experience described on p. 44 under the heading. The opportunity Grasped.

2.1. Which of these is most appropriate to the climate of science you would like for your school? Or perhaps you would prefer a balance ? What balanced combination of these four science activities would serve your needs?

2.2. Perhaps you will want to ask yourself the same questions about the activities, Winning Participation. The Science Shy; and The Science Prone.

NOTES

1. E. Bright Wilson, Jr., *in Introduction to Scientific Research*, McGraw-Hill, N.Y., 1952. Experimental design is discussed extensively in this book.
2. National Society for the Study of Education, *Forty-Ninth Yearbook*, Part I. *Learning and instruction*, U. of Chicogo Press, Chicago, 1950, p. 124.
3. *Ibid.*, pp. 115-16.
4. L. Klopfer and F.G. Watson, "Using Historical Materials in High School Science Teaching," *The Science Teacher*, *24*, 264, 1957.
5. Harvard U. Press, Cambridge, various publication dates; list prices vary from about one to two dollars.
6. Other useful examples of cases are available in, I.B. Cohen, *Science, Servant of Man*, Little, Brown Boston, 1948, and W.I.B. Beveridge, *The Art of Scientific Investigation*, rev., ed., W.W. Norton, N.Y., 1957.
7. F. Sherwood Taylor, *A Short History of Science and Scientific Thought*, W.W. Norton, N.Y., 1949, pp. 206-07.
8. Jan Ingen-Housz, London, 1779.
9. G.G. Simpson, G.S. Pittendrigh, and L.H. Tiffany, *Life, in Introduction to Biology*, Harcurt, Brace, N.Y., 1957, pp. 18-19.
10. Bertrand Russell, *Mysticism and Logic*, Barnes and Noble, N.Y., 1954, p. 14.
11. Carl Sandburg, *The People, Yes*, Harcourt, Brace, N. Y., 1936.
12. L. S. Kubic, "Some Unsolved Problems of the Scientific Career." *American Scientist*, 42, 112, 1954.
13. R. D. MacCurdy, "Characteristics of Superior Science Students and Their Own Sub-Groups" *Science Education 40*, 1956, p. 3.

    P. F. Brandwein, *The Gifted Studeut as Future Scientist*, Harcourt Brace, N. Y., 1955.
14. Lewis Terman, *Scientists and Nonscientists in a Group of 800 Gifted Men.* Psychological Monographs, Vol. 68, No. 378, 1954: also "Are Scientists Different." *Scientific American*, 192, Jan. 1955 pp. 25-29.

3

Science Teachers

A note at the beginning: Ah, the good old days. Claude Coleman gives us one reminder[1] of some of these "good old days." He records the account of a "teacher" who had taught for fifty-one years; during this tenure his services included "911,527 blows with a cane: 124,010 with a rod: 20,989 with a ruler; 136,715 with the hand; 10,295 over the mouth; 7,905 boxes on the ear; 1,115,800 slaps on the head . . . " One can guess at this "teacher's" pattern.

Teaching is a personal invention and every teacher has his own pattern. But there are different kinds of inventions and patterns. In this chapter we are concerned with three general types, each with a distinctive flavour. These center around three identifiable approaches in the process of teaching.

Three Patterns

As one observes teaching throughout the country (and we have observed hundreds of excellent teachers, and unfortunately many poor ones), one is forced to look at the teaching process in the public school in historical, perhaps even evolutionary, perspective. An observer becomes conscious that perhaps schools, and the teaching processes practiced in them, exhibit an evolutionary sequence and that different schools are at different stage of evolutionary development.

We see the teaching process developing toward a recognizable goal: the introduction into school and classroom of methods and procedures which are the applied results of half a century of research in the psychology of learning.' Where teaching procedures of an individual teacher or school ignore what has been learned from research and observations on effective teaching, the more primitive stage of the evolutionary process toward effective teaching exists. The direction of this process is *from* authoritarian teaching, with the objective of transmitting the knowledges, skills, and attitudes derived from a traditional subject-matter-oriented curriculum in which success is measured in terms of grades, *toward* a type of teaching which is the cause, as well as the result, of mature planning among teachers, children, parents, and administrators in a continuous attempt to improve teaching and learning. In the latter case the curriculum is planned cooperatively by teachers and supervisors, sometimes with parental assistance; units and lessons are planned cooperatively by teachers and students. Growth of teachers and students, and of the community, is the objective. One thing must be made clear: at least as much subject matter is learned with the latter approach, when it is practiced with skill. This approach begins with the needs and interests of all concerned (students, teachers, and the members of the community) in the learning process, whereas the former considers a temporary recollection of facts as the major hallmark of educated men and women.

A caution and a generalization: We presently know of no one way of teaching, no one curriculum, no one philosophy which fits all classrooms situations, all communities, all subject areas, all teachers. Research to date yields no one method applicable to the teaching scene generally. Teaching is essentially an art based upon the psychological results available. Above all else, the teacher is a responsibte human being in a situation rife with human aspiration.

A Pattern of Domination (Mr. A.)

Mr. A. was considered a good teacher of chemistry in the school where he taught. His students generally scored well on the state's standard examination; in fact, they were among the highest scorers. He rarely had discipline problems. Although he never volunteered, or undertook on his initiative, the supervision of a science club, no one really minded, because "he did his jab." And in the large high school where he taught chemistry there were others who took on the extracurricular activities.

When he spoke at faculty meetings, he would sometimes say, "Unless you give daily homework you can't expect results. These students don't work the way we did."

And in department meetings, when problems of discipline were discussed, he would say. "I never have any trouble; my students do exactly as I tell them."

As a matter of fact they did. One day when Mr. A. was absent another teacher tried to teach the class. He manipulated a problem in chemistry (weight-weight) in such a way that the students could not use the approach they had been taught by Mr. A. In effect the students told the teacher. "Mr. A. never taught it to us this way. He wants us to the problems in these steps. We are marked off it we don't do them his way."[2]

What these students were saying was, in essence, that the were becoming set in their ways of solving chemistry problems. They were being forced to follow one routine, which was presented to them ready-made. Instead of thinking through each new situation, they expected it to be like the last standard example (and it was; Mr. A. saw to that), and reacted to it in the same way. This is "stimulus-response" psychology carried to a ridiculous extreme. By demanding the practice of his routiue, clever as it may have been, Mr. A. was *reducing* readiness to think, *reducing* readiness to reason. His students were discouraged, even prohibited from attempting to seek concepts through creative problem solving.

These were students carefully being "prepared" for college entrance examinations. The reports from far too many college' teachers are that such students were hardly being "prepared" for collegiate study. Otherwise, why would so many collegiate instructors assert that they would rather (yes, *rather*!) have students who had not studied the subject, say chemistry, in high school? This is a devasting remark to make. It is almost the maximum insult to high school teachers; for have they not spent their lives "preparing" students for these very college courses? Surely something is basically wrong."

In an effort to explore this serious assertion, one of us has proposed a hypothesis which many of the critical college teachers say is essentially what they observe and what provokes their unkind remarks. It runs like this:

The entering college freshman believes he knows something about a subject because he has "had" a course in it. Actually he has a speaking acquaintanceship with some information and a recall of some generalizations. However, this exposure has covered too many separate topics; it is woefully "thin" everywhere. The student does not know how any of the generalizations were developed, he does not know where they apply; and, more serious, he does not know where they do *not* apply. He is very proud to write "H_2O" in place of "water" and never knows that H_2O, the real pure chemical, is a rarity even in collegiate laboratories. He knows that $PV=k$, but he does not know that this is a theoretical notion applying only to an "ideal gas" and only when the temperature is held constant. He does not know that most gases will condense to a liquid when the pressure is raised to a few hundred atmosphers. He has not *solved* problems: he has only *done* them. He has not attained, or formed, concepts; he usually repeats what he has memorized. Later, in college, he is obliged to "unlearn" or "relearn" what he proudly believes he knows, before a more analytical and realistic study of the subject can be presented, that is, before he truly attains the concept. He is too ignorant to realize how little he knows. He has been trained to jump through a certain set of hoops, but unfortunately in the real world, and even in college, jumping through these particular hoops is not considered especially important.

The tragedy here is that Mr. A. and his many colleagues thought that they were "doing their job," but they never tried to analyze what "their job" was. They have the knowledge, the intelligence, and the inherent abilities to bring children to self-controlled learning, but this never seemed important to them.

Mr. A. would never have thought of it this way: he would have rejected any suggestion that he wasn't teaching students how to reason. In fact, he believed that hard work in learning resulted in improved thinking—the harder the better. Mr. A. believed firmly that science develops logical thought. He was not interested in the fact that, more than a quarter of a century ago, the appealing assumptions of faculty psychology had been discredited experimentally. In fact, he used to say that psychology was not a science but "quackers." What he had to say about *educational* psychology would not bear repeating.

Mr. A. operated on schedules. His work was laid out carefully in a complete schedule posted on the bulletin board. He covered the entire course of study; he did not tarry for the slow nor hurry for the fast. When, year after year, the slow fell far behind, they were admonished to "work harder." The abler students were admonished to "work for perfection": to memorize and memorize until they knew 100 per cent of 100 per cent of the course. Enrichment and acceleration, Mr. A. maintained, could wait for college, which was the place for it anyway.

And the exams! Every three days, a short ten-minute quiz: every week (Friday), a full-period quiz. Every four weeks, a monthly quiz. The quizzes were mainly of the short-answer form with many problems and were always concerned with chemical details:

"Describe the Solvay process using equations only."

"Balance the following ten equations."

"What volume of bromine will be prepared if 200 grams of sodium bromide are used?"

And the entire problem was wrong if it wasn't done to two decimal places.

If one visited Mr. A. when he was not conducting a quiz, one could see the students at work, Yesterday's lesson had ended with an assignment, usually of this sort;

"For tomorrow, Do all the questions and problems."

Next day for twenty minutes, students were sent to the board to answer the questions and problems. For a few minutes Mr. A. answered questions about the chapter. When he found, through the questions students failed to answer, that they did not understand a chemical process, he gave a brief lcture-demonstration. In addition, demonstrations were done at the beginning of each laboratory period (held once a week) to be certain the students knew how to set up the equipment and what results to expect. Furthermore, Mr. A. introduced each topic, e.g., Oxygen, Solutions. Hydrochloric Acid. Halogens, Bases, Earth Metals, etc., with a lecture-demonstration.

Students had little to complain about. They had elected chemistry. They were taught chemistry. If they did their work. they "passed"; if they didn't, they "failed." Grades were given as a strict mathematical average on the basis of Mr. A.'s tests. Mr. A.'s students always did well in the examinations which the school gave. They did not generally gather about him after class, or after the course was over. They were polite to him, however, and he was polite to them.

They learned chemistry: he taught chemistry.

To the reader: *Turn now to ihe list of characteristics in the teaching process Check those factors which characterize Mr. A.'s approach, Or, if you wish, wait until you have examined the patterns of Mr. M. and Mr. P.*

A Pattern of the Laissez-Faire (*Mr. M.*)

Mr. M. was quite young. As a student he had become dissatisfied with the rigid patterns of the teachers in a "college-oriented" school he had attended. As he used to tell it:

> What they did was as predictable as the rising and setting of the sun. A lecture was given for 30 minutes, than, questions were asked for 20 minutes on the previous day's work. Everyone knew what the lecture was to be; it followed the text and references (college freshman texts) pretty much.
>
> The questions were predictable; they covered the readings assigned. If you memorized fairly well, you were certain of a good grade. If you didn't recite well the instructor usually reminded you that you wouldn't do well on the College Board Examinations. I was bored, and so were most of the other students.

As a student teacher, Mr. M. had observed not only a high school teacher in science, but an elementary school teacher as well. He observed how the latter apparently let children do much as they pleased. yet they seemed to learn very well. (He wasn't experienced enough, as we shall see, to note that she *knew* the children very well and *planned* very carefully).

He had read widely and was much impressed by what he considered progressive education: to wit, emphasis on the growth of children, not primarily on subject matter. He determined to try a version of his own in science; and he did so, in general science.[3] He determined, in short, to let children plan the course pretty much on their own. He determined to exert very little control. After all, he thought, there were no College Boards in general science.

The result was not what he had expected.

He began (the very first time he met them) by asking his students controlled in ninth grade science, "What do you want to do this term?" He impressed on them that: they could do what they wanted; he was there only to help them; they could work in groups, individually, or as a class; the examinations were to be made up by the students (with his guidance).

The class was gratified. The students were of middle-class homes and well-behaved; besides Mr. M. was personable and had a "way" with students.

A class chairman was apponinted pro tempore and the students fell to suggesting topics for study. Within a few days they had the following list on the board—each topic ably defended by its protagonists.

Psychology	Why do we behave as we do
Botany	Animals
Chemistry	The Microscope
Astronomy	Television
Bacteria and Disease	Airplanes
Satellites	Electricity
Evolution	Life on Mars
Nuclear Energy	The Human Body
Antibiotics	Reproduction
Fish	Farming
	Pets

As Mr. M. tells it:

I thought, even if this were hodgepodge, it still indicated their interests, possibly even their needs. Soon, however. I wasn't so sure; I began to suspect that some of them were suggesting topics just to get attention; some of them to please me; and some just for heck of it.

Nevertheless. I was going to see it through. Besides, my department chairman said, "Go ahead; let's see what happens."

The chairman of the department suspected what would happen, but he also knew that Mr. M. had great promise and would learn from whatever happened. So he left Mr. M. on his own to a great extent, although he was ready to help if any real "discipline trouble" arose.

There were sensible students in the group. They saw that several of the topics really belonged together and they argued this well. (Thus "Antibiotics" was placed under "Bacteria and Disease"; "Life on Mars," under "Astronomy"; etc.) After discussing a few "Rules of Getting Along and Getting Work Done," the group divided itself into committees (except for the boy who wanted to work on "Fish") and began to work.

Each day, the groups met, working in allotted corners and other spaces of the room, and in the library when necessary. Mr. M. sat at his desk ready to help. And students, appointed as group chairmen or self-appointed, did come up, asking such questions as:

What books can we get on glands?

How can we get Ralph to work? He just sits around.

When can we report to the class?

How do you want us to write this up?

We want to take pictures. Can you get us firm? We'll bring the camera, (Mr. M. did break the red tape to get a special budget for his class.)

As the work went on, Mr. M. began to see things he wouldn't have accepted until he had tried his way; some he noted to his dismay, some to his elation, while others just baffled him. What he saw, or consciously recognized, he stated in the report he made to his chairman. Here is a partial list excerpted from that report:

Some good things happened.

(a) I really think the kids learned to work together

(although I really can't know whether that was the result of the method I used—no experimental control).

(b) The best students learned a good deal. When reports were made they listened carefully, they asked questions, they read further. So they did pretty well on the exams. The exams were fair; I saw to it that they covered the work done rather well.

(c) The kids learned to report; even some dull ones talked up.

(d) I really believe some of the best were stimulated to go on to science because they had a good deal of freedom.

Most things, I would say, were negative or indifferently poor:

(a) A few kids took over; in each committee there was usually one.

(b) Not all kids worked equally hard; about one-third of the class loafted.

(c) After three weeks of work this way most kids were pretty bored.

(d) When they gave reports (after three weeks of activity. the first group on "Astronomy" gave an encyclopedic report runuing for ten days) most kids were pretty boaring; they read their reports, used few visual aids, and didn't speak up. In short, they just weren't teachers. I don't know why I expected them to be.

(e) The poorest students were just ieft out.

(f) In reporting (and in their work generally) most of the kids didn't know what weight to give important or unimportant aspects of their work. They slighted principles and emphasized details; they used vocabularies of unnecessary difficulty; they didn't use the board: etc.

(g) Worst of all I began to worry that they had left our some pretty important topics. (What of next term?) They hadn't selected "Photosynthesis," for instance; they had missed "Lighting at Home," "Fire Extinguishers," etc.

So in the middle of the tern, when a particularly bad two days (very boaring) had just gone by, when Joan asked worriedly, "How are we going to do in our next year's work, Mr. M.? All my other friends are taking different topics from those we are." I suddenly took the bull by the horns and asked, "How would you like me to take over for the rest of the term?"

I still remember the look of bright relief on the faces of almost all the youngsters. The few who weren't relieved were probably wondering why I hadn't done this in the first place.

Mr. M. would have saved himself considerable anxiety if he had read the famous study of K. Lewin, R. Lippitt, and R.K. White, "Patterns of Aggressive Behaviour in Experimentally Created Social Climates."[4] This study is also reported with photography in Goodwin Watson, "What Are the Effects of a Democratic Atmosphere on Children?"[5] There are many other related studies available.[6] Perhaps even more valuable would have been a thoughtful reading of any modern text on teaching methods.[7]

To the reader: Now Mr. M.'s experience is noteworthy and very much like the experience of "Mr. Osborne" as reported by Cronbach.[8] If you do study Mr. Osborne's experience, you may want to check that account, as well as our Mr. M. against our list of statements underlying class management in science, in the "Excursion" at the end of this chapter.

It wouln be fine if we could say that Mr. M. developed his teaching pattern into one in which coordination and planning became part of his teaching approach. Mr. M., however, did not get the best possible help from his supervisor or from administrators, who apparently wanted a pattern like Mr. A.'s but beliieved in a "sink or swim" approach to teachers. Mr. M. left teaching.

A Pattern of the Democratically Planned (*Mr. P.*)

If Mr. M. had had appropriate help from his supervisor and had had additional training, he might have developed the pattern of Mr. P., Mr. P, had oscillated between a pattern of domination and one of laissez faire: neither satisfied him. The latter did not leave room for the kind of responsible guidance he thought a teacher should give his students; the former did not encourage personal responsibility and growth in goal forming and concept seeking, which he believed should be a part of every person's education.

What was Mr. P.'s approach to teaching?

He taught in a school with a normal population, that is, a population diversified in I.Q., reading score, and socio-economic background. The students were expected to take a standard state examination; most of the college bound would probably take, in addition, the College Board Examination. The school was also expected to show up well in these examinations.

Most teachers in the school thought the best way to help students achieve success in these examinations was to teach the way our Mr. A. did—with kindness, of course. (The only trouble they could find with our Mr. A.'s method was that he pushed too hard; he was too tough; a bit of kindness would have helped Mr. A., they thought; you after all, "he did his job.")

Mr. P. did not believe this at all; he knew that there had been advances in the psychology of approaching class management and learning; he wanted to put these into practice in his classroom. He proceeded to do so. Let us say at the outset that his students did very well in the examinations; they even elected more science in later grades; more than that, a number far in excess of that expected took to science as a career; and more than that, a good number took to Mr. P. as a father-surrogate.

When he met his class, Mr. P. began by asking, "If you could have planned this course, what would you have included in it? Or, to put it another way, what would you really like to know? What would you like to learn?"

In the discussion which followed several things happened, most of them predictable:

1. Some students began suggesting topics such as these, which were put on the board:
   Why do we behave the way we do?
   Why are like our parents? Why aren't we?
   Life in the past.
2. Other students began suggesting topics such as these, also put on the board:
   Botany
   Zoology
   Digestion
   Anatomy
   Nutrition
3. Still others, cautious, even prudent, asked, "Don't we have to take the state examination?" To which the answer was an unequivocal "yes." Then: "Shouldn't we take the topics which will prepare us for that examination?" And some added, "For the College Boards as well.") To this also the answer was an unequivocal "yes."

The questions and discussion were resolved into a pattern which might be stated like this: "What can we do this year which will accomplish three purposes:

1. To study those topics which will help us live better lives; things we need to know.

2. To study those topics which will help us do our best in the required examinations.

3. To study those topics which are interesting (hobbies, projects. etc.). These may even be individual topics such as 'Cancer in the Swordtail' (a project done at home)."

During the next week each topic was discussed by the entire class (and we mean discussed!). Its advantages and disadvantages were weighed in the light of the amount of class time available, its priority in helping improve the quality of living, its usefulness in helping pass examinations (state and College Board), its special interest to a few students.

We have watched such classes at work, and the students are clearly sensible. They even admonish each other on wasting time in repetitious discussion.

At the end of a week or ten days, a list of topics and their subdivisions, as developed by a committee of students interested in a special topic, was placed before the class. The topics were in three categories usually. And Mr. P. attests to the fact that free discussion and skillful questioning almost always result in these three categories. (Mr. P. uses the pattern of teaching: always questioning, refraining from giving information.)

Category A. Topics we should like to discuss in class. (Includes activities we should lik to do in class.)

1. How our bodies work (anatomy, physiology; includes dissection of frog, rat; cells studies).
2. Why we behave as we do (elementary psychology).
3. What we inherit (heredity, reproduction).
4. Our origins (evolution; also such activities as a survey of animal and plant kingdoms to show evolutionary relationships; includes field trips).

Category B. Topics we can read for ourselves. Permit one

class period every week for questions (planned by a committee with teacher's advice;

1. Plants and animals (classification).
2. Conservation (to be reviewed when field trips are conducted).
3. Nutrition (to be reviewed when physiology is discussed).

Category C. Individual and group projects and readings. To be done at home or during free periods.)

1. Preparing permanent microscope slides.
2. Raising tropical fish.
3. The embryology of the snail.
4. Culturing protozoa.
5. Studying heredity of the characteristic "dimpling."

With three different categories of topics to be considered, Mr. P. and the class agreed upon a working procedure. Topics in Category A were taken up sequentially in class, each being considered for some weeks or even months. Each topic was subdivided into four or five smaller components on which committees worked with the constant advice and encouragement of Mr. P. Thus the whole class was at one time working on the same general area, yet there was opportunity for individual special interests.

Topics in Category B were, as mentioned, read from standard texts and other references outside the class, but one period each week was reserved for questions. The scheduling of the reading and discussion was planned by a student committee, who knew that they would be taking the required examinations.

Category C involved aspects which usually grew out of Category A or B and which fed back results when reports were made or questions raised. That is, individual expertness was encouraged and recognized. Mr. P. was available to advise on these projects, to redirect efforts that were staled, or to expand, through questions, efforts that seemed to have run dry.

Mr. P. had by this time convinced the class that they and he, together, could plan the course successfully. He remembered that only his first few classes had floundered; he had learned to trust his students. One of the things he had learned was that the success of the work depended on class committees and time given for planning. And there was the rub. There was never enough time.

Finally he had hit on this plan.

He met once a week with the committee chairman who formed the Executive Committee. One week the meeting was early in the morning (before school); another week it was in the afternoon. Here ways of proceeding were ironed out. Here reports, directions, laboratory work, films, field trips, lessons in which Mr. P. took the lead, were planned and scheduled. The committee also examined the three available texts and selected one for major use. In this way Mr. P. met with committees in all his classes.

Each chairman scheduled meetings with his own group out of class: at home, during stady periods, during lunch period, etc. In any emergency, the committee could call on Mr. P. Somehow he managed to arrange meetings. (The reader will see that Mr. P. believed in extracurricular work.)

Each committee assigned one person to the class laboratory squad. This laboratory squad planned the materials necessary, ordered them from the science department's laboratory squad,[9] and saw to the distribution of the materials, their collection, and their return. (The department laboratory squad also saw to the preparation of materials and their maintenance and repair.)

What was Mr. P.'s part in the day-to-day work? It was that of a teacher whose concept of science was described in Section I Hence, he believed in permitting students to do their "damnedest with their minds, no holds barred." The reader will have noted that the course the students planned included *more* subject matter, not *less*, than would be found in a course within the pattern of teacher-domination. This is generally the result of a teaching approach which permits the students to help in the formulation of goals and objectives. The morale of the class, the motivation of the students, is high when the purposes of learning are clear and acceptable to them. These students worked vary hard.

Mr. P. taught often; planning by students doesn't mean abdication by the teaeher. When he found that a students report needed clarification, he asked questions which illuminated. When he thought that a topic was not fully covered, he made suggestions. When the topic was too difficult to be handled by a student, or group of students, he became a member of the committee and planned with them; he taught actively, usually in the pattern of questioning.

He was a guide and friend; he was firm and demanding, but not coercive; he was permissive. There was an absence of threat in his class; there was freedom to proceed with the business of learning, not freedom to disrupt learning. Is it strange that Mr. P. had few discipline problems? The students and he, as teacher, were engaged in a commonly accepted, commonly planned activity.

What of the poorer student—the science-shy student? Actually he had a much better opportunity in Mr. P.'s class than in Mr. A.'s or Mr. M.'s. In Mr. A.'s, he would have failed. In Mr. M.'s, he might not have foiled, but the could have gotten by with doing very little. He might even have become a first-class nuisance. In Mr. P.'s class the other students expected him to contribute in committee work; he prepared a report, and he reviewad in committee and in class. Since he had a stake in planning the work, there was greater likelihood that he would work to capacity. In fact, Mr. P.'s classes were known for their healthy learning atmosphere; there were few, if any, discipline problems.

What of the science-prone student? Clearly he had an opportunity to exert leadership; clearly he could extend his learning beyond normal expectations; clearly he could work in a group or work individually. No limitations were placed on him; individuality was not crushed, but encouraged. In fact, Mr. P.'s approach depends on enlisting the leadership of the science-prone student.

In short, Mr. P. had developed an approach which enabled him to meet the needs and interests of different individuals within a class. This he did by adapting to the situation in class the principles of learning, gleaned from recent investigations of the conditions most fruitful for group learning.

Furthermore, we know that Me. P.'s students did as well in the state examinations and College Boards as did other students in his school taught by the lecture method.

And Mr. P. was one of the most, if not the most, respected teacher in the school, one of those most liked by parents and students, and clearly appreciated by his colleagues and the administration. His colleagues called him a "teacher's teacher."

An Excursion into Appraising the Teacher as Classroom Leader

Our readers will no doubt have put themselves in the places of our Mr. A., Mr. M., and Mr. P. This would be, at least in part, a share. For we insist that, an individual's origins and development being what they are, teaching approaches remain personal inventions. Nevertheless, there are frames of reference; and those of us who respect excellence aspire to it.

To repeat, the three "types" we have described, all actual cases, fit into the prototypes, as they emanted from the famous

TABLE 3.1

Some Consequences of Three Teaching Patterns*

	Undirected	*Group control*	*Teacher control*
Pattern	Uncoordinated	Coordinated	Dominated
Teacher's role	Vague adviser	Mature, wise group leader	Task-setter, drill-master
Emphasis	Monetary interests of individual pupils; "fun"	Significant group problems developed to fullest extent possible with maturity of pupils	Recall and recognition of academic subject matter, practice of academic skills
Direction	Unguided, uncritical	Self-responsibility, learning how to learn	Success on examinations, collegiate admissions
Emotional security	Disturbing because of low accomplishment of vague tasks	May be frustrated unless group plans clearly	Relieves anxiety by set standards, if clearly stated; teacher assumes responsibility
Enjoyment	Enjoyable until anxiety appears	Enjoyable if group sees progress toward clearer goals	Individual enjoyment from compliance to set tasks, but not enjoyed by all
Effort and efficiency	Low persistence, wasteful	Self-directed toward chosen goals, continuous when leader is absent	Effective if group accept direction of effort and leadership of teacher,

			otherwise resistance and minimal rerponse
Study patterns	Haphazard	Often intensive, self-initiated toward clear goals	Highly organized by teacher, littil pupil initiative
Assignments	Few, vague	Group planned	Set by teacher
Learning of course materials	Not much evidence, probably quite ineffective	As good as teacher controlled classes; possibly superior for alltering attitudes; approach emphasizes processs of scientific inquiry	Averages as good as group-controlled classes on academic recall and application
Evaluation	Testing difficult, too diverse activities; some appraisal of individual efforts	Tests involving novel problems-situations; continual observation, reports	Major reliance on tests, both standardized and home-made
Type of student favoured	Self-reliant, previously motivated	Equal opportunity and encouragement for science shy, typical, and science prone	Academically inclined conforming, college bound

* Based in part on L.J. Cronbach, *Educational Psychology*, Harcourt, Brace, N.Y., 1954, esp., p. 459, and W.H. Burton, *The Guidance of Learning Activities*, 2nd ed., Appleton-Century-Crofts, N.Y., 1952, after lengthy discussion, esp., pp. 325-27.

study of Lewin, Lippitt. and White.[10] With regard to the development of the science-prone student, or the "scientifically gifted," Brandwein[11] found that a situation which favoured the democratic planning" approach in the science class was more truitful in developing these students than either the "authuritarian" or the "laissez-faire" approach. This is in accord with the findings of Lewin, Lippitt, and White, and others. Futhermore, observations throughout the country indicate quite clearly, almost beyond the shadow of a doubt, that in the general estimates of students, supervisors, administrators, and the community, the teachers whose approach in the classroom is centered around "democratic planning" are considered to be most effective. Sometimes the oft-used phrase "democratic" seems to lose its central meaning: a systems of checks and balances to avoid both the domination of an absolute monarch (Mr. A.? and the chaos of anarchy (Mr. M.?).

3-1. Where do you place yourself? Where do you place Mr. A., Mr. M., Mr. P., and other teachers you have known, on Table 3.1?

3-2. In your estimation which of the statements below apply most characteristically to Mr. A.'s teaching pattern? Mr. M.'s? Mr. P.'s? To your own—present or planned? Or do they not apply at all?

1. Some science teachers drive, even bully, their students to work.
2. Some science teachers have classes which come in ready to start work and to enjoy it.
3. Some science teachers are humorous and friendly, albeit firm and helpful.
4. Some science teachers are in almost continual conflict with their classes.
5. In some science classes keeping quiet is safer than volunteering an idea that may be rejected.

6. In some science classes admitting one's difficulties to the class results in friendly and fruitful help.
7. Some science teachers believe that pupils will not voluntarily engage in useful work.
8. Some science teacher believe that students will engage in fruitful work, especiaily if they take a useful part in planning the work and working toward an acceptable goal.
9. Some science teachers believe that students can be left to their own devices, and can thereby achieve worthwhile learning.
10. Some science teachers believe that subject matter is something to be "covered" before going on to something clse.
11. Some science teachers believe that classwork is part of the general goal of teaching—helping students to improve the quality of their lives by directing their work toward enduring purposes and processes.
12. Some science teachers believe that since group control is the usual pattern in adult life, it should be increasingly used in class management.
13. Some science tsachers believe that adolescents do not know what they want or can do, and must be told.
14. Some science teachers, in their teaching, really satisfy their own needs (for security, status, recognition) rather than those of their students.
15. Some science teachers, in their teaching, attempt to satisfy their students' needs.

Of course, it will be said, and rightly, that these statements apply to all teachers, not to science teachers only. But the success of the science teacher is not measured only by his success in teaching science per se: we repeat, he teaches *science to students.* Hence the nature of his attitudes and approaches to his class,and his management of its students are of paramount importance.

The success of the science teacher in teaching science cannot be greater than his success in managing his science class.

NOTES

1. Claude Coleman, "The Hickory Strick," *Bulletin, American Association of University Professors, 39*, 457, 1953.
2. This remark always reminds us of two very interesting studies; A.S. Luchins, *Mechan ilization of Problem Solving—The Effect of Einstelling*, Psychological Monographs, Vol. 54, No. 248, 1942, and Hv. M. Schroder and J.B. Rotter, "Rigidity as Learned Behaviour," *Journal of Experimental Psychology, 44*, September 1952, p. 141.
3. In a Midwestern city school, Population 2,400 students.
4. *Journal of Social Psychology, 10*, 271, 1939.
5. *Progrecsive Education, 17*, 336, 1940.
6. For example, H.J. Bingham. "The Relation of Certain Social Attitudes to School Environment," *Journal of Experimental Education, 9*, 187, 1940.

   ——R.B. Embree, Jr., "Children Compare Two Types of Education Experience," *School and Society, 50*, 319, 1939.

   ——L.T. Hopkins, "Seniors Survey the High School," *Teachers College Record, 42*, 116, Columbia U., 1940.
7. W.H. Bulton, *The Guidance of Learning Activities*, 2nd ed., Appelton-Century-Crofts, N.Y., 1952.

   ——B. Othanel Smith *et al.*, *Fundamentals of Curriculum Development*, rev. ed., World Book, N.Y., 1957.
8. L.J., Cronbach, *Functional Psychology*, Harcourt, Brace, N.Y., 1954, pp. 442-46.
9. A discussion of student squads may be found in the companion volumes: Morholt, Brandwein, and Joseph, *A Sourcebook for the Biological Sciences*, and Joseph, Brandwein, and Morholt, *A Sourcebook for the Physical Sciences*.
10. *Op cit.*, p. 77.
11. Paul F. Brandwein, *The Gitted Student, As Future Scientist*, Harcourt, Brace, N.Y., 1955, p. 51.

4

The Course in Chemistry

A note at the beginning.[1] It is still true, of course, that college entrance requirements exert a strong influence on the high school course in spite of the fact that only 20 per cent of our high school graduates enter college.[2] It is also true, however, that many colleges are accepting modified chemistry courses as an entrance requirement in lieu of the more usual college-preparatory chemistry course. As a result, more and more boys and girls are enrolling in applied chemistry courses.

Note the term "applied chemistry course". Applied to what. Isn't the "college entrance" course in chemistry "applied" to some purpose? Or are these terms "college entrance chemistry" and "applied chemistry" meaningless?

For over a century chemistry has been taught in the secondary schools of this country. At present about half a million pupils, most in the eleventh grade, are enrolled in chemistry each year. Evidently the present courses are successfull or not too unsuccessful; otherwise the enrollment would not be this high. Yet chemistry in public schools enrolled in 1956 only 34.6 per cent of the children in the eleventh grade.[3] As science teachers we are acutely aware of the importance of chemistry. Certainly we can see that chemistry should be signi ficant to a greater fraction of the students than now elect it. What factors act to depress the enrollment? How did the courses now being offered come into existence? How might they be changed to be more attractive to the science shy, the average, and the science prone?

Development in the Course of Chemistry

Chemistry in American high schools shows, like biology and physics,[5] the strong impact of early collegiate domination. In the earliest American secondary school, the Latin grammar school. no science whatever was taught. This first school was intended to prepare students to read Latin and perferably Greek also in preparation for college. Colleges then were a far cry from those of the present; their major concern was the preparation of young men for the ministry.

Later the academies, like that founded in 1750 by Benjamin Franklin, presented a more varied and practical course of study including some science. But there was still no chemistry until the beginning of the nineteenth century. Then when the establishment of the public high school had begun, and after Lavoisier and Priestly had furnished chemistry with its first important concepts and procedures, chemistry gradually appeared in the American high school. For the most part this earliest chemistry. like botany and zoology, was taught catechistically as a series of teacher-questions and student-answers enlivened with dramatic demonstrations: the valcano, and assorted odours. In the early 1800's chemistry was considered (for some reason) a science most fit for young ladies, and many of the "female academies" advertised "a good chemical apparatus." Others foresaw the industrial and agricultural importance of chemistry and urged its teaching as a means of avoiding the necessity of importing chemicals from Europe. But generally there was no central purpose to the new course either in the colleges or in the high schools.

During and after the Civil War chemistry became more popular, and laboratory work for both boys and girls was stressed. In part, this stemmed from the wide acceptance of Froebel's emphasis upon work with the hands as a form of creative expression, and in part, from the success of the chemical theory of Avogadro as clarified by Cannizzaro in 1859. Now there was something to teach and a way of going about teaching it. Seemingly with more enthusiasm than wisdom, the "discovery method", or heuristic approach, was hailed as the key to all science teaching. By learning how to discover in science, students would be able not only to observe better, but to do better. In chemistry, as in the other sciences, laboratory work was the panacea, and unattainable objectives were propounded.

In keeping with the spirit of the times. Harward College in 1886 added "laboratory chemistry,' to the courses on its admi ssions list for students seeking advanced standing. To provide guidance to schools offering such a course. Professor Josiah Cooke of Harward prepared a list of 83 experiments, later reduced to 60, in qualitative and quantitative chemistry. Students were expected to perform most of these in school and were individually tested in the laboratory when they came to Cambridge to apply for admisson to the college. This list of experiments was soon known as "The Pamphlet" and went through several revisions as the result of criticisms and school experience.

Cooke's "Pamphlet" focused attention on the purpose of chemistry in the schools. Many hailed it; texts were soon written to conform to the plan. Others, however, bitterly complained that it was too difficult, too quantitative, too deficient in descriptive and qualitative chemistry, too mathematical, and too theoretical. Actually, this course was too advanced and abstract for many of the pupils. A few, but far from all, other colleges followed Harvard in accepting a course of this type for advanced standing.

In the *Report of the Committee of Ten on the Reorganization of Secondary School Studies* in 1894,[6] the recommendation was that chemistry be accepted for admission to all collegegs. This recommendation was not rapidly accepted: relatively few colleges would accept chemistry as a course for admission until some years later: Harward in 1900, Yale in 1911.

In 1902 Alexander Smith[7] found that chemistry courses fell into three major types:

1. The discovery or heuristic, with strong emphasis upon laboratory work.
2. The theoretical, which dealt with the gas laws, atomic theory, and Avogadro's hypothesis.
3. The historic-systematic, which dealt with oxygen, hydrogen, water, etc.

Although collegiate emphasis was upon the second form, the prevailing form practiced in secondary schools was the third. Ten years later, in 1912, textbooks were found[8] to be organized around water, air, salt, and the periodic table.

Throughout this period the schism between schools and colleges had been developing. Two major factors were working. First, the rapid expansion of secondary schools had produced a quantitative demand for teachers far inexcess of the number of able new teachers available: as a result many classroom teachers were themselves unable to handle effectively the details recommended by the colleges. Second, and perhaps more important, was the expansion and change in the student body within the schools. Teachers realized that they were expected, under compulsory education laws and general public enthusiasm, to educate *all* American boys and girls. With a heterogeneous student body having diverse interests and abilities, the rigid high school chemistry course proposed by Cooke was no longer suitable; for many of these boys and girls were not certain of completing high school, or were not necessarily going to any college, or found the chemistry course too dfficult or uninteresting, or were not mentally equipped to master the many principles, facts, and quantitative aspects of the existing course.

In retrospect the initial college domination appears to have had considerable value. Chemistry was moved into a position of academic respectability. New, if unsealistic, standards of performance and teacher competence were defined. Yet, as the high schools passed from college preparatory to common schools, changes were difficult and the definition of a new center of concern—a basis for more effective courses—was not sharply defined or made operative. The struggle to reorient chemistry has been a long evolutionary one. It continues, with vestiges of almost every earlier course plant still in use.

Numerous attempts have been made to reorient and redesign the course in chemistry. One has been to use the basic ideas of the college-preparatory course as a frame with changes in sequence and emphasis. The knowledge of elements, compounds and mixtures, physical and chemical changes, water and solutions, acids. bases, and salts, as well as some typical fundamental chemical reactions, was used as the basis upon which other optional topics were developed.

Another attempt was to explore the ways in which the materials of chemistry could be used to further general education, the content which would be most appropriate, and the methods which would be most feasible. In 1947 the Committee for the Study of Science Education commented;[9]

> The secondary school course has usefulness for the purposes of general education......Accordingly, chemistry teaching should clarify the relation of chemistry to health, to vocational pursuits, and to the other aspects of living to which the subject matter of chemistry relates and to which it can contribute understanding.

Not only the purposes and methods, but also the organization of chemistry courses have come under scrutiny. We quote again from Jaffe:[10]

> Organization of topics in a chemistry course has also undergone some modification. The popularity of textbooks which divided the course into *tight* units is waning. The

effect of this artificial grouping of chemical knowledge resulted in mental indigestion for the pupil and in a strait jacket for the inexperienced or beginning teacher......The [tight] unit treatment seemed logical enough but created difficult learning situations. Today, the trend in organiza tion of subject matter follows the psychological order. Difficult concepts and hurdles are carefully interspersed throughout the course so that understanding is not permitted to tailspin or interest to fall too sharply.

Patterns of Present Courses in Chemistry

The College-Preparatory Course

How well does success in the college-preparatory course in the high school enable us to predict future success in college ehemistry? The answer is "very poorly". About all we can say is that a student who has difficulty with high school chemistry will have little chance of success in the college course. Even outstanding success in secondary school chemistry is no guarantee of collegiate success. The theory studied in high school is necessarily elementary; that in college is considerably deeper. The college course also beging to stress mathematical treatment, which may overwhelm the student who needed little more than a good memory and simple arithemetic for success in the high school course.

Despite all the reports and studies, the present high school college-preparatory course is markedly similar to that of the 1920's.[11] One need only compare the chemistry syllabuses of New York State, which have influenced those of many other states, in the 1920's with the revisions of 1938 and 1956 to note how much of the fundamental framework remains intact. The syllabuses have been expanded (rather than revised) through the inclusion of recent developments which sequire greater time and attention than did the few materials deleted. Teachers' complaints are frequently heard; increasingly their courses become a race against the cock, and lectures must be used to "cover" the materi al.

Table 4.1 shows three typical college-preparatory syllabuses and the outline of a widely used college text. Compare them for details and organization. Also, if practical, compare them with similar syllabuses and the outline of widely used college text. Compare them for details and organization. Also, if practical, compare them with similar syllabuses from twenty or thirty years ago.

The principles course

While chemistry and all other sciences abound with principles, it does not follow that a chemistry course is necessarily a "principle course." This will depend largely on the content

TABLE 4.1

College-Preparatory Courses in Chemistry

High school (West)	*High school (East)*
History of chemistry	Introduction
Oxygen	Solutions and water
Hydrogen	Oxygen and hydrogen
Atomic theory	Atomic structure
Formulas and equations	Chemical nomenclature, formulas, equations, and problems
Sodium	Periodic table, metals, nonmetals, and inert elements
Sodium compounds	The halogens and their compounds
Chlorine	Sodium and calcium compounds
Problems (Avogadro's hypothesis)	Ionization
Sulfur	Sulfur and its compounds
Ionization	Nitrogen and its compounds
Reactions—reversible and non-reversible	Carbon and its oxides
Nitrogen	Nuclear energy
Halogens	Organic chemistry
Carbon	Metallurgy
Calcium and its compounds	Principles of reaction
Metals	
Important mineral substances	
Activity of metals, electrochemical series	
Compounds of carbon	

(*Contd.*)

TABLE 4.1 (*Contd.*)

High school (Midwest)	*College text*
Oxygen	The background of chemistry
Hydrogen	Measurements and methods
Water	The commonest element: oxygen
The structure of water	Another gaseous element: hydrogen
Formulas and equations	The nature of gases
Chemical calculations	Concerning atoms and molecules
Acids, bases, and salts	The compounds of hydrogen and oxygen
The halogens	The atmosphere and its constituents
Sodium, potassium, and compounds	The particles atoms are made of
Sulfur	The structure of atom
Nitrogen (The atmosphere)	The formation of compounds
Carbon and compounds	The release of energy
Fuels	The nature of the solid state
Calcium and its compounds	Solutions, their nature, properties, and laws
Metals	Some important complex ions
Iron and steel	Reactions depending on the mobility of electrons and of protons
Copper, zinc, tin, and lead	Speed and balance in chemical reactions
Aluminium and magnesium	The extremes of the periodic system
Mercury, silver, and other metals	Chemicals from salt and sea water
Colloids	Sulfur; base of modern industry
Organic chemistry	Phosphorus and fertilizers
Textiles, dyes, and plastics	Metallurgy; iron and aluminium
Foods, drugs, and cosmetics	Magnesium and calcium; hard water
Nuclear chemistry	Carbon and silicon, glass and ceramics
	A bit about organic chemistry
	The rest of the periodic system

*Note that many college courses do not follow this pattern, nor do all the college textbooks; see, for example, L. Pauling, *General chemistry*, 2nd ed., W.H. Freeman & Co., San Francisco, 1953.

selected and the methodology used to teach it. While newer statements of aims and objectives are putting more emphasis on principles, too often in chemistry the principles, are taught as isolated items not applied or developed through further work (e.g., the laws of definite and multiple proportion). Also, the interrelation among principles frequently is not established (e.g., is not applied in the the principle "reactions go to completion" study of industrial processess and the use of catalysts). Usually principles seem to be derived from the topically organized course illustrated in Table 4.1.

A different course in chemistry

As in other areas in science, there is a cutting edge, a point at which advanced thinking about the course of study, as well as experimentation, is taking place in the teaching of chemistry.

One such course, a very interesting one, is described here:[12]

> In much that has been written in recent years concerning the effective relationships between high school and college chemistry courses, two comments recur:
>
> (1) care should be taken to avoid wasteful repetition between the two courses, and
> (2) performance in freshman college chemistry appears to be little influenced by whether or not the student has had high school chemistry. This latter statement seems to imply that the standard high school chemistry course is ineffective as a basis for more advanced work. Furthermore, it borders on a gratuitous insult in the face of the fact that most college chemistry students say they found their original interest in chemistry during their years in high school .....
>
> At a conference sponsored by the Division of Chemical Education of the American Chemical Society and the Crown-

TABLE 4.2

Proposed Course Outline for High School Chemistry Based on Chemical Bonds as the Central Theme

I. Introduction
  A. Metric system
II. Elements and Atoms
  A. Laws of chemical combination
  B. Atomic weights and symbols
  C. Atomic structure
    1. Electrons
    2. Electronic forces; coulombic, exchange
    3. Atomic numbers; protons and neutrons
    4. Periodic table
III. Chemical Bonds—Discontinuity of Chemical Change
  A. Bond types; ionic, covalent, metallic
  B. Physical properties of substances
    1. Gases
      (a) Gas laws
      (b) Kinetic molecular theory
    2. Liquids
    3. Solids
      (a) Crystals, e.g., diamond, sugar, sodium chloride
  C. Physical transformation and temperature
    1. Gas to liquid
    2. Liquid to solid
    3. Relation of mass to properties
    4. Relation of transformations to bond types
    4. Classification of matter and physical transformations
      (a) Mixtures
      (b) Solutions
      (c) Compounds
      (d) Elements
    6. Purification procedures
  D. Discontinuities between elements and compounds
IV. Chemical Change and Covalent Chemical Bonds
  A. Reactive systems go to unreactive system
    1. Inert gases
    2. Reactivity and structure
  B. Methane, hydrogen, chlorine, hydrogen chloride
    1. Physical properties
    2. Substitution reactions
      (a) Formulas

(b) Equations
(c) Calculations
3. Chloromethanes

C. Oxygen, water, and carbon dioxide
1. Combustion
2. Chemical energy

D. Chemical geometry

V. Chemical Change; Metallic and Ionic Bonds

A. Atomic structure of metals
B. Oxidation and reduction (metals plus nometals yield ions)
C. NaCl, $MgCl_2$, KCl, MgO
1. Physical properties
2. Simple chemistry
D. Electrolysis to produce Na, Cl_2, Mg
1. Main chemistry of electrolysis

VI. Periodic Table

VII. Hydrogen, Chlorine, Hydrogen Chloride

A. Relative attraction for electrons, e.g. stabilities of NaH NaCl, and HCl
B. Polar covalent bonds
C. Properties of HCI

VIII. Properties of H_2O

A. Physics properties
B. Reaction with HCl
C. Reaction with Na

IX. Acids and Bases

A. Stoichiometry
B. Titration

X. Nitrogen and NH_3 System

A. Equilibrium

XI. Polyatomic Ions

A. Oxidation of NH_3 to yield NO_3
B. Sulfuric acid

XIII. Bonds Between Like Atoms

A. Carbon chains
B. Multiple bonds
C. Functional groups

---

*Inventions in science courses

Zellerbach Foundation held at Reed College, Portland, Oregon, in June, 1957, a group of high school and college teachers again discussed the integration of high school and college courses. It was agreed that repetition is a most useful aid in learning. The problem is one of judgement as to the material to be included at each level. The group at Reed College did not arrive at an answer to this problem, but one proposal emerged from the discussions which offers a way to move forward. It was agreed that a good high school chemistry course ought to have a quality of intellectual integrity that can be communicated to the student and that this could be achieved by having a focus toward which most of the discussion could be directed. If a course for high school students could be devised with a central theme less broad than the whole of chemistry, but including the major paths by which a chemist proceeds in his dealings with chemical phenomena, then it ought to be possible to produce a reasond argument for the topics to be included or excluded, the order of presentation, and the points at which individual variation might most readily be introduced.

A major differentiating aspect between chemistry and and other branches of natural philosophy is the concept of chemical bonds. Indeed, the making and breaking of these ties between atoms is chemistry. Our proposal is that "Chemical Bonds" is the logical central theme.** for a meaningful high school course. It is a theme large enough to include a great amount of descriptive chemistry and at the same time to serve as a guide to the items which can best be included in the course itself [see Table 4.2].

The Applied Chemistry Course

Compulsory education has brought to the high school a considerable number of students who are nonacademically minded. At best, the high school course will be terminal for these students; from their ranks come the many who will drop out of school before graduation. These pupils are not equipped to handle the more demanding features of the typical (college preparatory) high school chemistry course.

The coarse in applied chemistry was designed to present these students with a knowledge of chemistry that would achieve some of the broader objectives of general education. Typically, such courses devote about one-half to two-thirds of a semester to "basic" chemistry (from the concept of elements, compounds, and mixtures to a study of acids, bases, and salts). Abstract topics such as atomic structure or chemical shorthand are treated ligghtly, in keeping with the students inability to digest material of this nature. The groundwork having been laid, the balance of the year is devoted to a study of broad topics of industrial applications, such as fire and fuels, textiles, dyes ann bleaches, household cleaning materials of construction, and drugs and cosmetics. An example of a syllabus of this nature is shown in Table 4.3.

TABLE 4.3

A Syllabus in Applied Chemistry*

First Term, Class as a whole	*Second Term, Reorganization into Optional Groups*
	Optional Group A
I. The work of the chemist	I. Fuels and combustion
II. Nature of matter	II. Metals
III. Solution	III. Alloys
IV. Water	IV. Conservation of natural resources
V. Electrical nature of matter	*Optional Group B*
VI. Formulas and equation writing	I. Textiles
VII. Chemical arithmatic	II. Dyeing
VIII. Acids, bases, and salts	III. Cleaning
	Optional Group C
	I. Drugs
	II. Cosmetics
	III. Insecticides and repellents
	IV. Photography

* *Applied Chemistry for High School Students*, Curriculum Bulletin No. 2, Board of Education, City of New York, 1948-49 Series.

The Functional Course

While dyes, cosmetics, bleaches, and so forth, are important applications, of chemistry, we are naive to assume that the student inevitably has a burning interest in them. Those who defend the inclusion of such materials in a course cannot escape the fact that whatever reasons we advance for the importance of these materials are teachers' reasons. So far as the student is concerned, the reasons are tenuous. After all, of what value to him is a knowledge of these materials? In what may can he see a *use* for the knowledge? How do they relate the student to his environment, both local and national, in a significant manner? How will this knowledge influence his later behaviour?

Some school systems have striven toward a course designed to emphasize the vales of chemistry to the student as an individual, a member of the community, and a member of the nation. Such courses, designed to answer the questions above, are usually called "functional courses," a typical study, of which is presented in Table 4.4. These courses attempt to get closure to the student's "needs and interests."[13]

The Honour Course

The changing high school population focused our attention on the problems of the nonacademically minded pupil As is often typical in human relations, the good and superior students were "taken for granted." After all, they posed no obvious problems and were hence left to themselves. But the long-time result has been one of neglect; their talents were left unconsidered and unexplored. In recent years, it has been more useful to hear the question. "What are we doing for our brighter students?" Many schemes in answer to this question have to do with newer administrative procedures in the school as a whole, by way of honour classes and honour schools within a school. Where numbers permit, as in large cities, special schools may be set up for the talented with admission being by examination only. For chemistry in particular, there

TABLE 4.4

A Course in Functional Chemistry (Denver)*

The introductory uuits (12 weeks)*

I. Why study chemistry

II. Importance of chemistry in modern life

III. The fundamentals of chemistry

The functional unit (26 weeks)**

IV. Chemistry of the individual
- A. The chemistry of water in the body
- B. The chemistry of food within the body
- C. The interdependence of plant and animal life
- D. The chemistry of glands of internal secretion
- E. The chemistry of bacteria and disease
- F. The chemistry of drugs and medicines
- G. The chemistry of clothing
- H. The chemistry of cosmetics
- I. Chemical hobbies
- J. Vocations related to chemistry

V. Chemistry of the home
- A. Chemistro of cooking
- B. Chemistry of cooking utensils
- C. Chemistry of tableware
- D. Chemistry of fuels
- E. Chemistry of refrigeration
- F. Chemistry of sanitation
- G. Medical services in the home
- H. Chemistry of construction and construction materials
- I. Chemistry of paints, varnishes, and lacquers
- J. Chemistry involved in landscaping and gardening

VI. Chemistry of the community and the nation
- A. Chemistry of water purification
- B. Chemistry of the manufacture and distribution of foods
- C. Chemistry of sewage disposal
- D. Chemistry of purification of air
- E. Chemistry of industries

* Maurice Ahrens, Norris Bush, and Ray Easley, in *Science in General Education*, ed. by V.T. Thayer et al., Communication on Secondary School Curriculum, Progressive Education Association, D. Appleton Century, N.Y., 1938, pp. 465-75. This outline served as the basis for the text *Living Chemistry* (and its laboratory manual) published by Cinn.

** These three introductory units are devoted to the minimum of the fundamentals of chemistry needed for the functional units: matter, energy, structure of matter, formulas, oxygen and hydrogen (very brief), solutions and ionization, equations, acids, bases, salts.

*** Each of these lists of problems is not considered completely in each class, and no student studies all the problems taken up within the class group.

has developed the secheme of providining an advanced course for the brighter students. This is equivalent to first-year college chemistry. (The topics and treatment are those of a freshman course; a college text is employed.)

The pendulum seems to have come full cycle after almost a century, The early Harvard proposals, which affected chemistry teaching for almost half a century, prescribed a course which enabled the student to enter college with advanced standing in chemistry. There is an important distinction, though: this course was then given to *all* students; today we do the same thing, but *only with those select few* who are capable of, and interested in, such an undertaking.

Famous for Developing a Course in Chemistry

Objectives as a Frame of Reference

In 1925 Powers[14] summarized the objectives of high school chemistry teaching stated in many sources. The list was too long to be covered adequately within one year. Some of the objectives bowed to the "disciplinary value" of chemistry, while others were vague and without meaning. A committee writing later on this situation for the *Thirty-First Yearbook* of the NSSE reduced this long list to a few broad objectives:[15]

1. Pupils in high school chemistry courses should develop understandings of those fundamental concepts, major ideas, laws, or principles of chemistry that will enable them better to interpret natural phenomena, common applications of chemical principles, and industrial applications and uses of the principles of chemistry.

2. Pupils in high school chemistry classes should learn to use the process of effective thinking, problem solving, and techniques of study that are best adapted to the solution of problems within the field of chemistry, specially those which most often present themselves in daily life.

3. Pupils in chemistry . . . should develop those attitudes toward the facts and principles of chemistry and toward the methods of investigation employed in the field [of chemistry] that will serve as guides in their use of chemical facts and principles and methods of problem solving.

These objectives are essentially of, by, and for chemistry; they are rooted in chemistry as a subject distinct from other sciences and from life in general. They look to the results of the course in chemistry as being applicable only to chemical situations. These objectives, stated in 1932, seem too narrow today. Yet an examination of textbooks and courses of study will indicate that they are still being used widely.

But chemistry taught in the high school is clearly a part of general education and it has significance for all. If students are asked to name something in the home or among their clothes or a medicine, which the chemist has not improved, they will difficulties. Clearly chemistry has relevance to our life and living. Just as elearly the chemistry course in intent and content could provide a new insight into our modern civilization and into the impact science has upon how he live and what we live with. Pertinent objectives are those which students see as part of their lives. Why are these so rarely operating in classrooms?

Like many significant questions, this is easier asked than answered. Yet several separate factors are probably working. Often the teacher is not sure what his intent really is. Is he a chemist teaching? How can he resolve his interest in the subject area and the advice of distinguished chemists with the realities of the pupils in his classes? To what extent can he escape this dilemma by throwing the responsibility upon some examination for which he must "prepare" the students? Here we squarely encounter the conflict which arose in the early decades of the century. If we were obliged to "take a stand," we would prefer a teacher who thoroughly knows his subject but whose allegiance is to the students. He can use his knowledge to help them learnt as we shall see.

In eomments upon this failure of science teachers to accept and apply the objectives of general education, the Harvard Report states: [6]

> The teacher is not always clear whether he is engaged in special or general education, what proportion of his effort is to be spent on coverage and in being factually up-to-date, to what degree he is training for manipulative skill, and so on. From the point of view of general education, we are interested in these things not primarily for their own sake, but as they fit into an integrated intellectual structure. Science instruction in general education should be characterized by board integrative elements—the comparison of science with other modes of thought, the comparison and contrast of the individual sciences with each other, the relation of science with its own past and with general human history, and of science with the problems of human society. These are the areas in which science can make a lasting contribution to the general education of all students. Unfortunately, these areas are slighted most often in modern teaching . . . because the teacher is engaged in continuous struggle to encompass the subject matter. How is he, then, to deal with extra things—the critical examination, history, literature, and general cultural content of his subject? It is of course true that as *extra* things

> these aspects of scientific instruction should be impossible. *But they are not extra things—they are the very stuff of science in general education.* Once it is clear that he is engaged in general rather than special education, these are the things which must be emphasized, and to an increasing degree as the student matures. . . The *integrative element* is the student's own mode of life and his personal relation to the immediate environment.

Note, please, that this was written in 1954; does it not apply to the present?

Conceptual Schemes as a Frame of Reference

Objectives alone do not underpin a course. A courses has its framework in conceptual schemes as well. Chemistry, like all courses with a bona fide area of development, has concepts and principles which are peculiar to it. To mention a few:

> The behaviour of chemical elements can be predicted from a knowledge of atomic structure.
>
> Atomic structure provides an arrangement of the elements which shows them to exist in families rather than as individual elements.
>
> Chemical changes follow quantitative laws.

It is thus possible, if one wishes, to organize the course on the basis of broad conceptual schemes such as these.

Needs and Interests as a Frame of Reference

Anyone who has taken a course in chemistry knows how its study could fulfil certain needs and interests of young people. Here are some "problems" which relate directly to students' lives and will involve the normal content of the chemistry course in a way significant to students:

1. How can we avoid the danger of spontaneous combustion in the home? (Expand further to deal with

methods of fighting the possible types of fire in the home.)

2. What typical household chemicals may be dangerous to use? (We shall certainly cover a multitude of facts in answering this problem.)
3. What may be done at home to soften "hard" water?
4. How can we assure a safe water supply?
5. How can chemistry help us in the disposal of waste?
6. How can useful materials be obtained from farm waste?
7. How can chemistry provide for our needs for metals in the face of shrinking ore deposits?
8. How does chemistry help in improving our health? Our food supply?

Note that even these few problems relate the student to his home, his community, his nation, and his environment.

But these "needs and interests" may seem subject-centered rather than needs-centered. Somehow "needs-centered" involves rather personal choices. In a chemistry course in which one of us taught, the students chose as a central problem "The Analysis of Body Tissue." For the first half of the term, students planned to examine certain of the tissues, particularly blood, under the micro-scope. Then, using various texts in biochemistry as reference, they analyzed blood for its various constituents, using a gravimetric and volumetric methods.

The students were enjoying the course hugely (it had been going on for a month and a half) when one student, more concerned with the facts of life than with the facts of biochemistry, asked, "Does this course prepare us for college entrance examinations and the Regents Examination?"[17] The answer was, of course, "Not if it's a typical examination." To this answer students responded variously, but there was quite a bit of concern. The problem was solved, courageously, in this way.

Students and teacher agreed that even though it was not possible to orient this course toward the examination, some preparation for the examination was still desirable. They agreed in analyzing blood, they had learned the language and tools of chemistry. For instance, they had learned to write formulas, do quantitative problems, make solutions of the desired strength, and so forth. They further agreed that even with the best of intentions the entire standard course could not be covered. Hence, this plan was worked out:

1. Students studied past examinations. They worked out a plan of reading which would enable them to "cover" the necessary material. One day a week was spent in quizzing each other (with the teacher's help) on standard formulas, equations, preparations, and so forth. When a standard demonstration to illustrate the material under study was required, it was done by students or the teacher.
2. Meanwhile, they continued their work in biochemistry, and then went into organic chemistry (particularly esterification).

The students were successful in the state Regents Examination. As far as could be determined, students in this course did as well as other students of equal I.Q., reading score, and mathematics score who took the standard course. This no longer surprises us, although it did at first. Students who are vitally interested generally read more, do more, and learn more.

Note again that planning with students does not result in lowering "standards," but results actually in inereasing the vigorous applicatiun of the students. The teacher does not abdicate; he remains a teacher. This means raising the standard; this means that students work harder and learn more.

But what if a student had *not* asked, "Does this course prepare us for the college entrance and the Regents examinations?" In our experience, in years of planning with students (in classes where there are boys and girls who plan to go to college), this question has *always* been asked. Where there is an absence of threat, students do ask sensible questions and arrive at sensible answers—with the help of the teacher, of course. The teacher, we repeat, does not abdicate; he is *the* guide to learning. When he is such a guide, more subject matter is lcarned than when he is prescriptive; such is our experience.

And what of special personal interests? Crystallography? Crime detection? To learn firsthand how chemistry is used in the detection of crime, a student must become acquainted with a considerable body of facts about chemical analysis. If he is attracted by the interesting possibilities of electroplating or wants to know how a plastic is synthesized, he will soon be acquiring much specialized information on his own initiative, and judging the significance of this information as it applies to his problem. He will be practicing the art of the scientist.

In another school, a student askcd during the first week what was left after the oxygen had been evolved from $KClO_3$. It "should" be KCl, but was it? The student was encouraged to find out, and over the next weeks was involved in techniques and learnings which "normally" came later in the course. Because he wanted to answer a question, he willingly learned much that is generally considered difficult. Since not every student will be interested in the same topic, such projects provide for individual interests. They can be carried out during, before, or after class, either at home or at school.

Units as a Frame of Reference

More than any other dcvice. the unit has been used as a base for orgnization of the chemistry course. Clearly units can be organized around conceptual schemes, problems, needs and interests, or utility in our society. In chemistry, units have been organized in topics (oxygen), conceptual schemes (periodic table), or fields (metallurgy). There seems to be no rhyme, if indeed there be reason, for this organization.

For instance, a unit on metallurgy may be based not on a single conceptual scheme or a "need and interest" or even a problem, but on the utility of metals in our society. Nevertheless, in terms of the task of maintaining the student's interest, a unit on metallurgy whieh stresses the general principles in obtaining, using, and protecting all metals clearly surpasses the disjointed treatment in which the metals are given individual study, each isolated in its chapter.

TABLE 4.5

A Course Based on Large Understandings*

I. Chemistry in the present world-introductory

II. A chemical view of matter

III. Some common chemical elements-common structure

IV. The world of water and solutions

V. Important classes of chemical compounds

VI. The families of elements
   A. The halogens
   B. Sulphur
   C. Nitrogen, including the atmosphere

VII. Chemical theories and their applications

VIII. The world of metals

IX. Carbon and its compounds
   A. The oxides
   B. Petroleum chemistry
   C. Organic chemistry

* In each unit the necessary skills (writing formulas and equations, problems, etc.) are taught.

Whether or not the frame of reference is in conceptual schemes or personal interests, the basic organization necessarily includes a large body of subject matter, the mastery of which results in understanding of some large *principle or function*. Usually, skills and attitudes ore stressed in relation to principles and problems. Nevertheless, the units of a chemistry course often do not have a parallel construction.

Table 4.5 is an example of one course in which the units do have parallel structure based on large understandings or functions.

Special Considerations in Teaching Chemistry

The Flavour of the Course

Chemistry, like each of the sciences, has its distinctive flavour. A visitor to a chemistry classroom or laboratory will likely find the teacher or students:

Decomposing an oxide.

Preparing a gas: hydrogen, oxygen, bromine.

Purifying water.

Discovering equivalent weights, e.g., of magnesium.

Finding the weight of 22.4 litres of oxygen.

Neutralizing acids and bases.

Preparing ammonia, nitric acid, hydrogen sulfide, sulfuric acid.

Studying destructive distillation.

Preparing an ester.

Determining an "unknown."

These operations and many others are the essence of chemistry. Also involved are:

A certain kind of mathematics.

A shorthand.

An extensive technical vocabulary.

Skills in designing apparatus and in manipulating equipment. These create the odours and colours, the "feel" of chemistry; they belong to it and to no other study. Because there is so much to *do* in chemistry, it can be treat fun.

Chemistry and Social Problems

Because chemists have been so successful in understanding

and maneuvering the components of the physical world, many fundamental changes in the social domain result. Probably these are not "controversial issue," as some applications of biology are considered, but they are social problems resulting from chemistry which must be solved in part with the relevant knowledge of chemistry. Among many are these:

1. Man has learned how to transmute the elements and release enormous energies. What shall we do with this new skill?

2. Nuclear experimentation and power plants are already putting into the air disturbing amounts of radioactive material which "fall out." How shall be deal with this? How significant are the amounts already in the air? What are the political and social implications, on a world-wide scale, of this "fall-out"?

3. The mineral resources of the world are finite and unevenly distributed. In the United States live less than one-tenth of the world's population on less than one-fifteenth of the land mass. Yet we use one-half of the world's mineral resources, iron, aluminium, oil, rubber, and so forth. What are the consequences and responsibilities of this rapid use? What problems does this pose for the near future? For certain necessary materials we are a "have not" nation (wolfram, manganese, etc.). What consequences does this have on international relations? What is the likelihood of substitutes being found?

4. We are using fossil fuels coal and especially oil, at a terrific rate. Yet the demands for power rise continually How will these he met? Who will make the decisions and finance the operations?

5. By 1970 our national population will be close to 200,000,000. Water is already in short supply in some areas, in part because it is used so extensively in manufacturing processes. Food for our people and possibly for the exploding populations elsewhere in the

world must be grown on the same limited and well-used land areas. What can be done, by whom, and at what cost to supply more water and food?

Are questions such as those above properly a part of the chemistry course? We believe they are, if we wish to have future citizens who comprehend the significance of chemistry in the world today. Yet they may be "side-stepped" in a teacher-dominated course limited to "chemists" chemistry and the "ivory laboratory." Such limitation will provide more time for technical details, but the only knowledge of the social implications of chemistry that the student will have is the haphazard knowledge he would have picked up had he not taken the chemistry course.

The Problem of Safety

Safety in the laboratory, where acids, bases, poisonous materials. glass-ware, flames, and high and low pressures are necessarily used continuously depends upon both knowledge and attitudes. Certainly the student needs to know what materials are hazardous and what to do if an accident occurs. But knowledge is not enough. In many chemistry laboratories we have seen, the attitude of the students was far too causal. They were only "messing around." usually to "prove" some result they already knew and did not care about anyway. More careful work will result as soon as the student *wants* his own answer, and a good one, to his question. Then he will be tidy and precise. An approach which involves predictions from knowledge will identify the dangerous materials and reactions; e.g., will this reaction be strongly exothermic? If so, what would you expect to occur? What precautions should you take? This will also reduce the chances that students will, as happens too often, "pour things together to see what will happen" and have a ghastly accident. Certainly there is no single answer to this difficult question, but the attitude and approach of the instructor can make a significant effect upon the atmosphere of experimentation.

Theory vs. Application in Chemistry

From the beginning of the century the relative attention to theory, application, and descriptive chemistry has been argued. No over-all answer can be created, for the intent of the teacher and pupils will vary among schools. Yet chemistry is a scienee, which means that the multitude of specific reactions can be organized and explained through a relatively few generalizations. In some instances, a "law of behaviour" can be formulated directly from the observations. In other cases, some postulated property of atoms or molecules (valence, electric charge, binding energy, elastic collisions) must be introduced as an intermediary in reaching an explanation. The success of chemistry in predicting what will occur results from these generalizations. To omit most of them for purely descriptive discussion of the wonders of modern technology prevents the student from comprehending the significance of the descriptive information and the basis on which technologies' with multi-million dollar investments, can be created.

Historical Sequence in Chemistry

Every science has a long and distinguished history. This is the story of how we have the knowledge now available. A course in chemistry which uses the historical approach would begin with the Greeks. But for a course in chemistry reflecling its use in the modern world, we would accept the concept of atoms as a starting point surely the idea of atoms will not be a novelty to the students. Then gradually questions arise about how we know when we can rely upon the existing generalizations. This is the time to repeat, perhaps, in the modern context, the classical investigations of the past (history) upon which our current knowledge rests. That is, the student needs to be aware of "how he knows" and "how well he knows." At appropriate times the struggles, achievements, and disappointments of former chemists can be used through brief case histories.[13] The purpose of this is not to teach the history of chemistry, but to use the past to illuminate the present. A student may then see how present problems are like those encountered by chemists in the past.

The Value of Laboratory Work and Demonstrations

After the heyday of enthusiasm for laboratory work, for which preposterous claims had been made, thoughtful people began to ask what effects laboratory work had upon children's learning. In part this was motivated by the "enormous expansion" of enrollments in public high schools (from 519,000 in 1900, to 2,200,000 in 1920, to 4,399,000 in 1930). The resulting strain upon building facilities, especially laboratories for individual experimentation. was severe and expenses were high, The possibilities of group laboratory work and of teacher demonstrations were explored.

During the second and third decades of this century many experiments were made to reveal the relative merits of these different procedures. At that time educational experimentation, which was just beginning, severely suffered from "faulty and inadequate experimental and statistical techniques and the lack of reliable and valid objective tests for measuring the instructional outcomes."[19]

The nature of these insufficiencies is exhibited well in one of the later and larger studies: that by Carpenter which involved over a thousand pupils in 34 classes in 23 schools in 14 states. We quote his major conclusion, supported by the data despite the primitive statistical treatment:[20]

> The results of this experiment point to the conclusion that the majority of students in high school laboratory chemistry classes, taught by the demonstration method, succeeded as well as when they performed the experiment individually. *if success is measured by instruments which measure the same abilities as are measured by these tests*, namely, specific information and ability to think in terms of chemistry.

For this to have meaning we must examine the tests he gave the pupils. As he states, they heavily involved recall of information. The "ability to think in terms of chemistry" is not so evident. The test items were based on "ten introductory experiments as found in one of [the] most popular chemistry laboratory manuals: Brownlee and others, *Laboratory Exercises*, Allyn and Bacon" (edition and cited). A perusal of these test items reveals that a competent student could answer them completely on the basis of a textbook knowledge *without either* demonstration or laboratory work.[21] Little wonder then that he did not find large differences between the results of presenting the phenomena to the students in the two different ways.

Among the many studies, probably the "classic" is that by Horton.[22] Not only did he devise standard paper-and-pencil tests which required more knowledge of "doing" or "performing," but, in addition, he devised a test of laboratory manipulative skills, appraised both through written tests and direct tests in the laboratory. After extensive experimentation, he concluded:[23]

> No reliable results appear in the testing by the ordinary written examinations—neither by the Regents nor by the school test.

This is not surprising in view of the type of item and ability involved in such tests. Also he concluded:[24]

(A) The order of preference of the methods studied, in the light of all the outcomes measured, appeared to be:

1. Individual laboratory *without directions*, the so-called "problem method."
2. Individual laboratory work following directions, but with these directions consciously generalized.
3. Individual laboratory work following directions from a manual.
4. Demonstrations of all experiments by the teacher.

(B) For success in the ordinary written test there was little preference; no method showed a superiority amounting to certainty . . . . We were left with a choice of believing that:

1. Written tests are invalid for detecting differences in methods of laboratory work.
2. The methods of doing laboratory work are not determining factors of success in written tests.
3. Both these statements may be true.

[Not more than one-fifth of the student's time in the course was devoted to the laboratory work.]

(C) There were. nevertheless, differences in the outcome of the different methods of using the laboratory. This appeared to be confirmed by all the non-written tests proposed. These differences were found in ability.

1. To manipulate apparatus.
2. To make experimentation involving use of apparatus.
3, To solve perplexities or projects involving use of chemical facts in laboratory situations.

He also found that 87 per cent of the students preferred individual laboratory work in demonstrations and commented:

From such a decisive preference for individual work we may judge that it serves at least the purpose of self-activity, motivation, and maintaining interest. On these grounds alone, the expenditure of time and money may be justified... It appears that we have overemphasized individual work as a means of acquiring information and of merely understanding chemistry. On the other hand, there seem to be important outcomes from individual work measurable and attainable be suitable methods which may make it more, rather than less, important.

All this only re-emphasizes the basic question of the desired outcomes of a chemistry course. Do we wish the pupils to attain primarily a static knowledge of chemical facts? If so, laboratory work, and perhaps even demonstrations, will not be essential. Tests will deal with recall of static knowledge of the type in texts. Incidentally, enrollments will be exceedingly low.) Do we wish the pupils to see chemistry as a dynamic science, as an example of how men struggle to describe the behaviours of the world and to make sense out of them? If so, laboratory worg will have a central place and will much more than exercises repeated from a manual. The issue is just that sharp.

The questions have remained unresolved. In 1946 the committee reviewing this topic for the *Forty-Sixth Yearbook* of the of the National Society for the Study of Education wrote:[25]

> It is regrettable that in a majority of science classes in which demonstration and individual experiments are performed. the chief, if not the sole, function served by these activities is to verify facts and principles already learned. The experiment is commonly postponed until the pupil has found out from consulting the text or from classroom discussion what the experimental results ought to be. Then the experiments are performed so that the pupils may verify what they already know. As a result, it is common for the pupils to engage in such undesirable practices as "making the answer come out right," and telling what "ought to have happened" instead of what actually happened.
>
> Performing demonstration or individual experiments merely for the purpose of verifying facts already known is rarely, if ever, justified. The primary purpose of experimenting is to secure evidence which may reveal answers to problems. In order to effect this purpose, the inductive method should be used in nearly all cases, that is, the laboratory work should precede, not follow, the classroom discussion of a topic or principle.
>
> Under this plan, the pupil needs the same careful

direction for experimenting that he will require in any other case; he needs to be given or to work out for himself with the teacher's help, the procedures to follow in securing an answer to a problem. But everything possible should be done to encourage him not to ascertain by other means what the answer should be. . . . After he has arrived at the best answer to his problem that he can reasonably be expected to obtain it is proper for him to consult the textbooks and references to find which are the correct results, as obtained by skilled scientists working under ideal conditions.

Such practice is ideal for the teaching of the scientific method and for developing scientific attitudes. The practice of carrying on experiments for the mere purpose of verification often emphasizes the antithesis of scientific method.

The thirteen years which elapsed should have brought a shift in emphasis in the objectives of the laboratory work in chemistry. But, to our knowledge. this shift exists in only a minority of classrooms. In a great majority of chemistry classes the statement which begins this section still applies: "It is regrettable thut in a majority of science classes...the chief, if not the sole, function served by these activities [demonstrations and individual experiments] is to verify facts and principles already learned."

This opinion was accepted as describing the present situation by chemistry teachers with whom we discussed the problem in sixteen Science. Teacher Institutes held during the summers of 1956 and 1957. It is also what we have found when discussing the situation with teachers throughout the country. In general, we find that the syllabus topics stated in Table 4.1 are used (in one sequence or another) in the majority of classrooms in this country. Similarly, we find that the topics shown in Table 4.6 constitute the laboratory exercises generally used throughout the nation--almost to the sequence.

Note that we say "laboratory exercises"; these are done by

TABLE 4.6

Typical Laboratory Exercises in Chemistry

1. Physical and chemical changes
2. Elements, compounds, and mixtures
3. Matter in chemical change
4. Decomposition of an oxide
5. Oxygen
6. Hydrogen
7. Displacement of hydrogen from water
8. Water and solids
9. Purification of water
10. Equivalent weight of magnesium
11. Weight of 22.4 litres of oxygen
12. Ionization
13. The electrolysis of an electrolyte
14. Neutralization
15. Reactions to completion; reversible reactions
16. Hydrolysis
17. Chlorine
18. Hydrogen chloride
19. Bromine
20. Iodine and a fluorine compound
21. Nitrogen (the atmosphere)
22. Ammonia
23. Nitric acid
24. Oxides of nitrogen
25. Sulfur
26. Hydrogen sulfide
27. Sulfuric acid
28. Carbon
29. Destructive distillation
30. Carbon dioxide
31. Fermentation
32. Esters

(*Contd.*)

33. Soap-making
34, Foods
35. Textiles
36. Relative activity of metals
37. Metallurgy
38, Qualitative analysis
39. Oxidation reduction
40. Aluminium
41. Calcium, magnesium, and their compounds
42. Hard water
43. Determination of an unknown

---

students after they have the information. Information is not gathered inductively through the experience. The laboratory experience is not in search of meaning; it is merely an exercise in laboratory manipulation, sometimes indifferently performed.

This is a most regrettable commentary about the role of laboratory work when all sciencc teachers are aware of the central place of experimentation in science. If laboratory work is handled in the manner described just above, the abilities of children to select and design equipment, to predict from principles, to operate carefully and accurately, to observe closely, to appraise results, to search for improved techniques and equipmental design, to apply statistical analysis, to describe data graphically and algebraicaily, and to interpret data will not be realized. Further research on the benefits of laboratory work would normally include these attributes as central. Yet in terms of current laboratory practices, we would expect to find little or no gain in the critical scientific skills.

But basic in the current misuse of laboratory time and facilities in the clear observation that children *enjoy* laboratory work, especially when their effects have some personal significance. A major revolution in science teaching with a sharp rise in enrollments can be forecast if the laboratory is used in the sprit of science to inquire, to test predictions, and to wrestle with reality.

An Execution Into Developing One's Own Course in Chemistry

We are forccd to the conclusion that chemistry teaching, as defined by the topics taught in classroom and laboratory, is fairly standardized throughout the country. The course given is that delineated in Table 4.1; the laboratory course given is that in Table 4.6. This standardization and orientation is unlike general science and biology. What does this mean?

Perhaps this: Chemistry teachers have devised an invention which is satisfactory. It meets the needs of boys and girls, the various communities, the schools, the colleges. Furthermore, chemistry teachers are generally satisfied with this invention.

Or this: Chemistry teachers are teaching a course which fits very few young people. This might account, in part, for the relatively low enrollment in a very interesting subject.

Or this: Change in chemistry teaching is slower than in other areas. In this aspect, it is like physics teaching.

Or this: Chemistry teachers are satisfied to teach only a select group. They do not, in general, believe that all young people should have the kind of chemistry which fits their individual goals.

13.1. One teacher we know developed an entirely different course in chemistry in this way. He decided that he wanted to prepare the youngsters for that bugaboo. The College Entrance Board Examination,[26] and yet to develop a course according to what was known about modern methods of teaching, the method of intelligence, and concept formation.

This is what he did; it is similar to Mr. P.'s plan described in Chapter 4. He realized that the major obstacle to teaching in a leisurely manner in a course so crammed full of facts was the lack of time. How to get time?

(a) He analyzed current textbooks and found them all very similar in content. But he also discovered that certain portions of the information need not be taught in the classroom; students could read and understand them: the history of chemistry. certain aspects of metallurgy. soap-making, etc. He did not, however, avoid *using* aspects of these topics, particularly history, to develop the way the chemist used the method of intelligence, and metallurgy and soap making to illustrate the aspects of technology.

He therefore developed an agreement with the students. He prepared a mimeographed sheet listing the topics for which they were responsible when tests were given. He also listed the dates by which these topics were to be read. (These were flexible deadlines and were changed as the course developed.)

He decided that in class he would deal only with the "difficult" topics, those which had a mathematical context, dealt with laws, or dealt with conceptual schemes. All descriptive chemistry he left to the students.

(b) He allowed one period (or occasionally two) per week for claring up difficulties which came out of the students readings and performing the standard laboratory and and classroom demonstrations.

(c) This left to the classroom and the laboratory such problems which developed from the very nature of chemistry, e.g.,

How do you identify an unknown ion in solution? (Much laboratory time was spent on analysis,)

How do you know that MnO_2 is a catalyst? (Experiments on catalysis were done in the laboratory.)

How do you know that weight-weight relationships really exist, e.g., in the reaction,

$$NaCl + AgNO_3 \rightarrow AgCl + NaNO_3?$$

(Careful experiments were done over a period of a week.)

(d) As a test of whether students "knew" chemistry, he used the type of test developed, as well as various American Chemical Society tests. He found that, when taught this way, his class did as well in terms of knowledge and skill gained as they did when he was (as he put it) "bound by the rat-race of the textbook-and-laboratory course."

There were three most important results:

1. Those students who were going on to college did well in the College Entrance Board Examinations.
2. Those who were science shy did as well, in his opinion, as they would have done in the course he used to give.
3. The enrollment in chemistry tripled over four years.

4.2. In order to determine what is going on in chemistry courses throughout the country you might want to gather various courses of study. The approach we have found useful is described at the end of. The Course in Biology.

4-3. For laboratory experiments, demonstrations, field work, projects, films, etc., its chemistry you might want to examine:

J. Richardson and G.P. Cahoon, *Meterials and Methods in in Teaching Physical and General Science*. McGraw-Hill, N.Y., 1952.

A. Joseph, P.F. Brandwein, and E. Morholt, *Teaching High School Science: A Source-book for the Physihal Sciences*, Harcourt, Brace. N.Y., 1959, a companion to this Volume.

NOTES

1. Quotation from Bernard Jaffe. "Trends in High School Chemistry," *National Association of Secondary School Principals Bulletin*, Jan. 1953, pp. 67-75.
2. This percentage has been rising and is now closer to 30.
3. K.E. Brown, *Offerings and Enrollments in Science and Mathematics in Public High Schools*, U.S. Office of Education Pamphlet No. 120, U.S. Government Printing Office Wasington, D.C., 1957. For a fuller discussion of trends in enrollment in high school science courses, see our Chapter 12. The Course in Biology, and Chapter 21, Supply and Demand in Science.
4. This section is based in part on S. Rosen. "The Rise of High School Chemistry in America (to 1920)." *Journal of Chemical Education*, 33, 627, 1957, and "A History of Science Teaching in the American Public High Schools, 1820-1920," unpublished Ph.D. thesis. Harward U. Cambridge, 1955.
5. See Chapters 12 and 14.
6. American Book, N.Y., 1894.
7. Alexander Smith and E.H. Hall, *The Teaching of Chemistry and Physics in the Secondary School*, Longmans, Green. N.Y., 1902, p. 19.
8. S. Rosen, *A History of Science Teaching*.
9. National Society for the Study of Education, *Forty-Sixth Yearbook* Part 1, *Science Education in American Schools*, U. of Chicago Pres, Chicago, 1947, p. 199.
10. B. Jaffe, op. cit., p. 71.
11. For various statements see the following in the *Journal of Chemical Education*; "Report of the New England Assn. of Chemistry Teachers' Committee on College Entrance Examinations," *16*, 46. 1939.

    "The Chemistry Examination of the CEEB." John T. Tate. *17*, 443, 1940, and *18*, 444, 1941.

    "A Minimum Syllabus for a College-Preparatory Course in Chemistry." John C. Hogg. *27*, 46, 1950, and M.M. Whitten, *34*, 307, 1957.
12. Laurence F. Strong and M. Kent Wilson, "Chemical Bonds; A Central Theme for High School Chemistry," *Journal of Chemical Education*, *35*, 56, 1958.

13, For brief descriptions of other courses of this type, see A. Laton and S. R. Powers, *New Directions in Science Teaching*, McGraw-Hill, N. Y., 1949.

14. S. R. Powers, *A Diagnostic Study of the Subject Matter of High School Chemistry*, Contributions to Education, No. 149, Bureau of Publications, Teachers College, Columbia U., N. Y., 1924, also, "Achievement in High School Chemistry—An Examination of Subject Matter," *School Science and Mathematics*, 25, 52, 1925.

15. National Society for the Study of Education, *Thirty-Five Yearbook*, Part I, *A Programme for Science Teaching*, U. of Chicago Press, Chicago, 1932, p. 259.

16. *General Education in a Free Society*, Harvard U. Press, Cambridge, 1945, p. 155; italics added.

17. Regents Examinations are the state-wide examinations of the Board of Regents of New York State; they reflect the state syllabus.

18. Leo Klopter and F.G. Watson, "the usr of case histories in science teaching" in *Science Teaching Through Problem Solving*, Nat'l Sci. Teachers Assn., 1956, mimeographed, p. 40; see also *The Science Teacher 24*, 264, Nov. 1957.

19. NSSE, *Thirty-First Yearbonk op. cit.*, p. 97.

20. W.W. Carpenter, *Certain Phases of the Administration of High School Chemistry*, Contributions to Education, No. 191, Bureau of Publications, Teachers College, Columbia U., N.Y., 1925, p. 45; Italics added.

21. Carpenter did not use his "control groups" (no contact at all with the phenomena) in his analysis.

22. R.E. Horton, *Measurable Outcomes of Individual Laboratory Work in High School Chemistry*, Contributions to Education, No. 303, Bureau of Publications, Teachers College Columbia U., N.Y., 1928.

23. *Ibid.*, p. 84.

24. *Ibid.*, pp. 99-102.

25. *Op. Cit.*, pp. 51-52.

26. Taken in 1954-55 by 14,933 children out of some 180,000 enrolled in chemistry courses, *Fifty-fourth Aunual Peport of the Director*, C.E.E.B., N.*Y.*, 1956, p. 62.

5

Secondary School Chemistry

A Look at Chemistry Teaching

Myron Eldermon was returning from a meeting of the Village Council at Wexford that warm April evening. This session of the council had been stimulating. The village councilmen had been discussing the advisability of installing a municipal water-softening system, and the exchange of thought and opinion was heated as the need and the cost of making such on investment were argued. It was all the more stimulating because of the fact that it might go into a special election to be held at a rather early date. As he walked toward home, Mr. Elderman saw in the situation many potentialities for his science teaching at the high school. Several students in his chemistry class were closely identified with the problem. He had observed that Bill Arthur's father was on the Council and was very much in favour of installing a water-softening system. On the other hand, Irene Dupre's father was not on the Council, but he was present and spoke vigorously against the proposal. It was obvious that he and many others who were against the idea held this position because they had installed home water softeners. It was their position that those who wanted to have water softeners should provide their own; in any event, it was foolish to soften all the water used in the town, particularly with as much as water.

In a few days the class would be closing the present unit of work. Myron Elderman would have to plan fast if the next unit was to be concerned with water softening. On arriving home he would make a few notes before retiring. These notes would incorporate some of the major ideas that come to him. But as he sat down to list these ideas he found that the list was growing longer and longer, and he worked late that evening. He could foresee not only the usual preparation and comparison of temporarily and permanently hard water, but there were many other investigations that could grow out of this problem which confronted the village fathers. His list included many such as these.

Jost How Hard is Wexford Water?

There had been agreement in the Council meeting that it was very hard water, but disagreement as to how hard it really was. Myron Elderman could sce both qualitative studies in this and comparison with the water in other communities as well.

What Makes Water Hard?

It was evident that some who had attended the Council meeting and spoke knew so little about the courses of hardness that the need for attention to it in the school curriculum was clear.

What is the best Method of Softening Hard Water?

There had been much disagreement on this point. It was evident that it was entirely too soon for the village fathers to consider the type of installation that would be desirable.

There was immediate interest of the class in the proposal that they study the water softening problem in Wexford. First, he asked the class about then feelings and the questions that they had on the subject of water softening. There were many; some were pertinent, a few were not. There ideas were listed on the chalkboard as they were given. Ted Smith and Irene

Dupre volunteered to organize the questions and problems for the class and the next day. Mr. Elderman helped them prepare the organization and added a few ideas. When the list was complete they were duplicated for class use.

The class time the next day was used in studying the reorganized proposals and in planning investigations. Laboratory work that seemed desirable for everyone was agreed upon and planned. In this instance, it included a study of soap-making, a comparison of industrially produced soop and a comparison of the hardness of water brought in from other communities. The proposal that water from various communities be compared was made by Alberta Parker, who came by school bus from Collinsville which was in a valley about eight miles distant. Keith Warren from Booth Township would bring a sample from that community. There was time left in the period for initial library reading in the science room, and Mr. Elderman listed other readings available in the school liberary. Overnight reading in the textbook was included.

Laboratory investigation began the next day, several of the students had prepared their plans overnight while others needed still more time to bring their plans to completion. It was obvious that some of the students had been talking about water softening with people in town. No doubt some of this had been stimulated by their parents. It developed that Jean Overbeck had been discussing the problem with Professor Clopp, of the chemistry department of the local college. As a result she suggested to Mr. Elderman, when he stopped by to discuss the apparatus she was assembling, that Professor Clapp might be asked to talk to the class, The proposal was placed before the class at the end of the hour, and everyone thought that it would be a good idea.

At the beginning of the class period next day, Bill Author came up excitedly with a proposal. "Why couldn't we prepare a summary of the results of our studies here and present it to the Council? It seems to me that we one learning more about this matter of water softening than many of the people who are arguing so much about it."

"It should like a good idea to me," was Mr. Elderman's reply. "Why dont we see what the class would like to do?"

There was immediate approval, both from those who seemed to be a favour of a municipal water softner and those were not. There was agreement that everyone should keep a close record of all that he learned and all that he thought would be appropriate to the study, so that the information might be brought together for the benefit of the people in the community.

During the next few days, various investigations went forward. Soap was prepared and tried. Water softened in home systems was brought to the school for testing. Samples from various communities were studied. One of the boys demonstrated the use of a Permutit working models of this softener because several of their homes were equipped with water softener which utilized the same principle.

In the meantime Mr. Elderman had located a filmstrip which described in some detail the ion exchange principle of water softening. The filmstrip was used by several of the students. Soon the reports of the various investigations were prepared. One report included an investigation of the hardness of water and its actual meaning. Another had to do with the effect of the boiler scale on the efficiency of water heaters. Jim Johnson, who had something of a flair for mathematics, calculated the approximate cost of boiler scale to the community. Arrangements had been made for Professor Clopp to come to the school and talk with the class about the problem of water softening. A committee of students headed by Jean Overbeck had talked with Mr. Clopp, told him of their investigations and asked him to help them with the problem of water softening in the local community. He was very happy to do so and used an entire class period. The students were greateful for his discussion and took copious notes.

The class selected a committee to draw together the materials from the various individuals and committees for the preparation of the final report. As fast as the reports come to the committees they were edited and summarized for the Village Council.

An Analysis of Activities

The teaching of chemistry can be a forward-looking investigation of everyday problems and phenomena, as it has been described in the foregoing example or it can be a meaningless memorization of facts and principles. The difference lies in part in the objectives of the teaching and in the activities provided to attempt the realization of the objectives.

In the situation deseribed, the teacher utilized a community problem of some significance. In their effort to contribute to this problem, the students participated in planning and organizing the study; they helped to determine what experiences would be common and required, and, as the investigation developed, what individual projects would be included Individual laboratory work was provided, including a study of materials (water) brought from their homes. There was reading and planning, and the use of a filmstrip. Water softening was integrated with other work in the school. A specialist from the community contributed to the study. Students reported in both oral and written forms.

From a chemical standpoint, the activities included analysis of quantitative and qualitative character. Industrial processes were reproduced in the laboratory, and industrial products were tested. Titrations were carried out, and there was a demonstration of the use of ion exchange in water softening. Further study would reveal other essential chemical operations.

Within the framework of such an investigation there are many potentially educative activities. Problems of significance in science involve an increasing array of experience for their solution. Only as the school provides the essential resources can the experience make its contribution to learning.

PLANNING FACILITIES FOR HIGH SCHOOL CHEMISTRY

Chemistry is a highly specialized division of the sciences, and the activities involved in its study tend to be correspondingly specialized. There is much of chemistry in everyday life and activity, and much of the provision for the experimental and experiential study of chemistry is relatively simple. However, the facilities required for student experimentation and investigation beyond the most elementary chemical phenomena are of such nature that careful planning is essential, Likewise, those facilities for demonstrating and for various specialized teaching procedures must be studied with a view to their contribution to the learning situation. In chemistry, as in other sciences, the major emphasis should be placed on the development of scientific thinking.

Experimentation and Observation

Over a period of many years a traditional pattern of laboratory "experiments" has been developed. To serve this pattern of activities rather conventionalized facilities have been provided. Provisions have been made for such operations as weighing and heating reagents, measuring the volumes of gases, filtering precipitates, dissolving solids, neutralizing acids, electrolyzing water, and burning various inflammable gases.

As the curriculum of the school is extended to encompass the needs of students and the needs of society, and as the fields of chemistry and chemical technology grow, the number and kind of student experiments and experiences will increase, Accompanying this increase, there is, in general, a greater complexity of experimentation. There has been a pattern of content and laboratory work considered suitable to high school chemistry as a preparation for college work in science. For the majority who do not attend college and for those who do not plan to specialize in scientific and technical fields the traditional patterns are considered by many to be too limiting.

Facilities for teaching chemistry should make provision for

the appropriate learning activities for all students, regardless of their vocatioaal or academic plans and interests. In addition to the provisions for conventional activities in chemistry, there should be facilities which provide opportunities for such activities as the preparation of plastics, identification of texttile fibers. production of paints, the removal of stains, the determination of the acidity or alkalinity of the soil, testing of foods for nutrients, the softening of hard water, constructing a lead storage cell, analyzing of fertilizers, and the production of alloys. On occasion these and other activities are the laboratory work for many or all of the members of the class; in other situations they are extended as individual or group projects or may even become research investigations.

It is important that flexible facilities be available to take care of the needs of individual experimentation. Two or three special cubicles or spaces provided with the necessary services make a valuable addition for the students who wish to set up special or individual experiments which can remain undisturbed.

Demonstrations are profitably employed by the teacher and the students in studying chemistry. Several studies have been made of the relative merits of demonstration and individual laboratory work: it is clear that some situations are served better by demonstrations than by other procedures. Students learn by demonstrating to others: certain apparatus and materials are expensive; some experiments are dengerous; in other instances it is more economical of time to employ a demonstration procedure. The essential factor to consider is effectiveness; true individual experimentation is necessary in many situations to bring about effective learning.

The facilities of the school should provide for a board range of experimental and demonstration activity. With a few of the major basic resources, the range of facilities can be extended as needs warrant the extension. As such times resources in other parts of the building and in the community can be made a part of the facilities for teaching chemistry.

Developmental Activities

Storage and experimental facilities far beyond those available in many schools are required when increased use in made of the project procedure in the teaching of chemistry. Such projects range from an investigation of short duration to one of several weeks or months; from individual investigation to one involving several students: and from such techniques as building and assembly to research activities. Individual projects may be the centre of class and laboratory study or they may be limited to times other than scheduled class time; in either case, they require storage space. Power tools and hand tools should be available. There should be a wide variety of chemicals and raw materials for construction activities. An abundance of reading material appropriate to the interests and concerns of the students should be provided.

Departmental work in chemistry involves not only the usual weighing, heating, filtering, burning and titrating but also various other operations of more involved nature. In addition to the familiar bending and drawing of glass tubing, there are occasions to make a glass T-tube, drill a hole through a sheet of glass, or cut a large glass tube. The use of Pyrex glass requires a supply of compressed air or oxygen for glassworking.

Chemistery teachers should know how to use common wood and metal working tools in order to direct project work. Equipment for demostrations and experiments can be devised and made by the teacher and students.

Display and Illustrating

The art of display and illustration is essential to the communication characteristic of a good learning situation. It is through effective display that a student can communicate to others something of what he has learned. Through display and illustration the effective teacher conveys his ideas to others or arranges a situation conducive to learning. Materials for producing illustrations and making display should be available for such activities. In this respect the fine arts laboratory should be a ready resource to the chemistry teacher. Posters. charts, tackboard displays and exhibits are valuable to the students and the tcacher. Their preparation is in itself an educational activity.

A darkroom adjacent to the chemistry laboratory permits students to take and develop photographs for use with their project work. Such photographs can be displayed, for the benefit of others. in a corridor display case. Exhibits of completed projects, collections, and other informative material should be displayed not only in the laboratory but also in a corridor display case.

The preparation of disply materials for snch community, intra and inter-school activities at School Science Night, Science Fairs, and Junior Academy of Science displays is a type of activity conducive to learning. Such activities trend to bring science into the lives of all students. Many of our future scientists are recognized in this way and can be encouraged to continue their investigations. A great variety of materials is required to encourage this developmental work.

Projection

The experiences essential for realization of the objectives of chemistry are so many and so varied that they cannot possibly be presented in their original form. It is necessary to use projected teaching aids to develop the most important of these. Facilities should be provided to make possible the projection of flat pictures, motion pictures, slides and slide films. In addition, the microprojector finds occasional use.

Projection equipmcnt is a traditional resource of the teacher. Facilitics for science teaching should provide also for student use of this equipment. Projected material is in itself a resource to the student, just as is a book, or a drawing he has made. The resources of the laboratory should provide for projection activit'es of students and teachers.

Reading, Writing and Discussion

A significant portion of the student's time in the chemistry room is spent in reading and related activities. Such activities consist of note-taking and writing the results of experiments, the taking of tests and writing of reports. Often class periods are lengthened and a portion of the time is used for directed study. Larger working surfaces are needed for books and writing supplies. Reading materials should be available in the laboratory either on a permanent basis or a loan arrangement witn the school library. Shelves for books, periodicals, and pamphlets make such material immediately available to the students. Tables and chairs are provided to encourage the students to stop and read.

The scicnce room or rooms should provide for discussion by the class as a whole, and by small groups. Flexible seating arrangements serve both general purposes, and tend to improve the quality of the discussion through the encouragements of informality. Club activities, also, are promoted through attention to the provision for such seating arrangements. The fixed tables of the chemistry laboratory reduce flexibility and if such activities are to be carried on more space must be provided.

Maintenance

Much laboratory time is consumed by maintenance activities ranging from repair of apparatus to washing glassware. Apparatus should be kept in good working order at all times and repairs should be made immediately. Many such repairs can be made in the laboratory if proper tools and workroom facilities are available. Delicate or complex equipment requires handling by specialists and must be sent out for maintenance.

The students in chemistry should learn the necessity for good housekeeping. This includes the proper handling of apparatus and reagents to avoid breakage and loss or danger by spillage. Both hot and cold water should be readily available at the wall sinks. The tables should be washed after use and broken glassware removed from the tables and floor.

Waste materials of a dangerous nature must be placed in suitable containers and arrangements made for the removal of such material from the room. The ever-present danger of fire in the chemistry laboratory requires that the students be protected by the presence of fire extinguishers and proper first and supplies.

General Aspects

The inherent qualities of chemicals are such that special attention should be given to the facilities used. Particular attention should be given to the corrosive effects of many substances and to provisions for ventilation.

Floors. It is important that laboratory floors be of a material that resists acids and othes corrosive materials which may be spilled. A quarry tile with narrow joints is one type of satisfactory floor. The chemistry laboratory may be floored with asphalt title if there is tittle use of petroleum products which disolve it. The asphalt tile, however, dries out and chips. Vinyl plastic gives promise of being a satisfactory material showing resistance to grease, oils, and alkalies as well as acids.

Decoration. The effects of corrosive vapours on paint is a factor in decorating the room. The paints used must no contain lead or chrome pigments if darkening is to be avoided. Water-base paints should not be used where washing will be necessary since such paints have low resistance to the washing compounds available for school maintenance. Among others, rubber emulsion paints show satisfactory properties for this type of use and they give evidence of being the most satisfactory wall finish, all factors considered.

Ventilation. The chemistry room commonly presents a most difficult ventilation problem. The release of obnoxious and/or corrosive vapors creates the need of special attention. The problem is reduced through attention to such procedures as the use of demonstrations where obnoxious gases are produced, utilization of semi-micro technique, and careful use of fume hoods. The use of small quantities of materials in student experiments helps to reduce the production of obnoxious gases and also conserves material. Even under the best of conditions, however, the need for good ventilation persists.

In planning for ventilation as the building is designed, care must be exercised that the air flow through chemistry classroom-laboratory is adequate. There must be no recirculation of air from the chemistry rooms into other parts of the school building.

Fume hoods with exhaust fans assist in keeping the classroom-laboratory free fram fumes. Ducts carrying such fumes should lead directly outside and must be made of, or lined with, non-corrodible material.

CLASSROOM AND LBBORATORY FACILITIES FOR HIGH SCHOOL CHEMISTRY

The design of the chemistry classroom-laboratory or of the laboratory alone should be related to the selection of its furnishings. Equipment cannot be fitted to the room; the room must be designed to accommodate the desired equipment. Many local factors, sucb as the size of laboratory, cost, and student requirements in relation to community needs must be considered in the selection of furnishings. The cost of furnishings must be considered in relation to their useful life.

Quality of Material

The arguments advanced in favour of wood and of metal furniture are given in Chapter II. Metal furniture may be undesirable in a laboratory which is to be used to both physics and chemistry because of its magnetic properties.

Corrosion is a problem peculiar to the chemistry laboratory and must be considered in the selection of furnishings. It is serious, particularly around hoods and drains where there is direct contact with corrosive chemicals. The materials used for table tops. sinks. and drains must be chemibaliy-resistant.

The Demonstration Desk

A well-equipped demonstration desk is essential in the classroom-laboratory and is quite desirable in a layout which provides a separate laboratory. The demonstration desk in the natural focus of attention and must be ample in size, with adequate accommodations and service facilities. For small classes a desk surface of 12 sq, It is satisfactory.

The demonstration desk should be equipped with several drawers and a cupboard. An open shelf to provide space for a tote box or temporary storage is desirable. A complete set of all the ordinary equipment necessary for demonstration use should be available in the storage space of the desk. It should also provide space for storage of ordinary classroom supplies such as paper, pencils, crayons, student records, and other materials.

The demonstration desk should have hot and cold running water, preferably with pantry cocks. Gas, if available, should be supplied by two or more gas cocks. If piped gas is not available, the demonstration desk and student desks can conveniently be supplied with gas from cylinders containing gaseous fuel liquefied under pressure. A large sink with a trap is essential. Several duplex electric outlets should be placed on the desk, with additional polarized outlets for direct current if it is available. Upright metal rods with a crossbar are very desirable especially if the room is used for both chemistry and physics,

Laboratory Desks

Movable student laboratory tables are satisfactory for many activities in science; however, the service facilities required for chemistry necessitate some minimum provision of fixed desks. Desks for both class and experimental purposes may be used (refer to Chapter II). Greater flexibility is possible if the student desks and sinks are placed around the walls; fewer sinks may serve if this arrangement is used.

Chemistry desks should be provided with gas, water, and electrical outlets. Both hot and cold water may be provided. The desks should be equipped with pantry cocks, a sink of chemical-resistant material, and drains and traps of lead or silicon iron. The choice of material for drains is governed somewhat by structural factors. Suggested sink materials are soapstone, chemical stoneware, or stainless steel.

Each student should be provided with a small drawer of sufficient size to hold the apparatus for his personal use as well as protective apron. Locking drawers built into the desks serve this purpose. The desks should have sufficient storage space to serve the anticipated total number of sections. In addition, some open free space should be available for each student to place books and other personal items while doing experimental work.

A wall sink for general use by students is desirable. It should be provided with a splashboard, double drainboard, and hot and cold water, A pegboard for draining glassware should be close by. A soap dispenser above the sink and a paper towel dispenser nearby are necessities. Cabinets under the sink store supplies such as paper towels, special cleansing agents, and brushes.

Fume Hoods

A fume hood with cabinet support and gas and electrical outlets is desirable for student use. A switch provides control of the exhaust fan; a pilot light in the circuit shows when the fan is operating. If running water is provided in a large fume hood, the area can be used for general experimental work.

The cost of installation of a fume hood is such that the advisability of the investment is subject to question. Various factors enter into the decision: in some states fume hoods are required by code; a fume hood is used only a few times each year; demonstrations using relatively small quantities of reagents can prevent the release of large quantities of obnoxious or harmful gases; semimicro laboratory procedures can be used by the students, thus producing very small quantities of gas.

There are other factors which have bearing on the question: in view of the desirability of providing for the continuous ventilation of certain stored chemicals, it is possible to provide for this in connection with the venting for the fume hood; it is possible to utilize a large fume hood as a space for ordinary laboratory aetivities, so that it is actually in use as much as other work areas.

Special ventilation of the room may be substituted for fume hoods but the safety values of the hoods can not be ignored. In view of this,, some educators ask for more than one fume hood in a laboratory, In some schools small down-draft hoods are used on the student tables. Such hoods prevent crowding at a single hood.

Display and Storage Facilities

Display facilities should be provided for at least two purposes:

1. to display within the classroom or the laboratory foı the chemistry students and,
2. to display in other parts of the building for the student body of the school.

A large amount of tackboard space should be available, as well a stand for charts or models, with suitable lighting. Such materials as student-made charts, flowsheets of chemical processes and cartoons can be used in a variety of ways in teaching.

The enrollment in chemistry and physics comprises only a small portion of the school population. The remainder of the students have tittle realization of the applications of many simple phenomena, There is real need for display cases in alcoves and corridors.

The school library often uses a display case and bulletin board to feature new books against an attractively arranged background. Such background material for books in science can be provided as part of the work of the chemistry students. Chemistry projects can be exhibited in the library to the advantage of the many students in the school.

An excellent display space can be provided by a case recessed in an interior wall. It should have a plate glass front on the corridor side and an access door on the inner wall. Adequate interior lighting should be supplied by indirect sources. A wall switch in the corridor permits the operation of exhibited devices requiring eletcric current.

The local science fair is a development which permits the exhibition of student project work for the school and community. Projects conceived and developed through the facilities of the laboratory are tangible evidences of achievement on the part of the students. Such activities require display space and facilities; these needs should be anticipated at the planning stage.

Storage requirements for apparatus can be partly met by using conventional storage cases, recessed in interior walls. These units in the classroom permit better supervision of storage by the instructor. Such units can be combined with comparable open shelving units for convenient storage of laboratory balances and other commonly used equipment. The same type of unit can be used to house the classroom library. Storage cases of great depth do not conveniently recess into walls; they can be placed along the walls only with the requirement of additional floor space and greater hazard in the room.

Those storage cases used for chemicals must be acid-resistant and if made of metal must be coated with acid-resistant material. Flammable chemicals should be stored in a steel cabinet.

A display and storage case with a removable horizontal

glass top should be provided. It can be placed in the classroom or in the laboratory or combined room. This type of case permits display or supplementary teaching materials, or student projects, and at the same time provides at least partial storage for these materials.

UTILITIES AND SERVICES

In planning classrooms and laboratories, it is essential that the type of equipment and services be selected so that the necessary floor and wall installations can be provided during the basic construction. The number of floor connections is determined by the choice of student desks. The cost of installation is related to the number of outlets for each of the services provided. While costs must always be considered, the controlling factor should be the desirable arrangement of the room. A given desk in a classroom-laboratory may accomodate 6 (or more) students for each set of connections but a room so equipped may be so inflexible that it cannot be used for any other purpose. A slightly great investment in service connections can provide a room suitable for various types of activity appropriate to chemistry classes and for other classes as well.

Water and Drain

Running water is a necessity. It is quite desirable to provide running hot water as well as cold. The hot water should be supplied from a water heating system of such capacity that peak demands of others do not interfere with its use in science. Both hot and cold water should be supplied to the demonstration desk, preparation room desk, and wall sink. Frequently, only cold water is supplied to the student desks: however, the availability of running hot water facilitates many operations. Cut-off valves should be placed in each supply line.

Sanitary drain lines from chemistry laboratories should be connected directly with the community's sewer system. If no municipal system is available, an independent system must be designed so that wastes from the laboratories do not pass into the septic tanks. In either case an acid proof drain should be used to carry the waste water from the laboratory sinks.

Gas

Gas should be supplied to the demonstration desk, preparation room desk, fume hoods, and to each student desk. Duplex outlets are advisable because two students can be supplied at the same time by one pipe fitting. Cut-off valves should be placed in each supply line and, in addition, it should be possible to shut-off the gas supply for the room with a single valve. Such valves should be of a type that can be locked, or should be located in a locked compartment.

Electricity

The electric lines should be designed to carry an amount of current necessary to supply all anticipated needs. Provision should also be made for special writing to supply special equipment, such as fume fans.

The writing for general purpose circuits, for lighting fixtures, the special spotlighting for the demonstration desk, the darkroom and the wall receptacles should provide for 30 amperes. Wall receptacles should be placed at convenient locations around the room.

A special outled should be located in the projection area. This should include a receptacle for the projector power cord, and a polarized receptacle for the speaker cord. This polarized receptacle should lead, through conduit, to the speaker near the screen at the front of the room.

The wiring for the demonstration desk and for the preparation room desk should provide for a minimum of 30 amperes. The demonstration desk should have a duplex receptacle of the pedestal type mounted on its surface; flush rsceptacles should be placed on the rail of the desk and on the side near the floor.

The electric circuits should provide a duplex receptacle for each four students. Separate fuses should be installed for each circuit. If the same room is to be used for chemistry and physics, it may be necessary to equip the student desk's with additional polarized receptacles to supply direct current from a battery or motor generator. If separate rooms are used direct current is supplied for the physics room, it may be desirable to equip the chemistry desks with polarized receptacles for direct current, or, as a minimum, to provide for direct current at the demonstration desk and at one or two other points in the room.

A separate circuit is needed for the display case built into the classroom wall. This circuit must supply current for lighting the case and for the operation of any electrical apparatus exhibited in the case. Wiring may also be provided to operate the display from a swith in the hall. This requires a special receptacle inside the case connected to the swith in the hall.

A switchboard is necessary for all a.c. and d.c. circuits. The fuse panel should be adjacent to the switchboard and readily accessible to teachers. Both the switched and fuse panel should be covered with locking doors. Rectifiers in the laboratory may be used to supply low voltage direct current.

EQUIPMENT AND SUPPLIES

The range of equipment and supplies of chemical nature needed for experimentation, demonstration, and project work is extensive, but without such resources the value of chemistry to the student is seriously impaired. While the ingenuity of the teacher and the students will go far in providing apparatus, many of the resources must be obtained by purchase because substitution is imppossible.

Equipment for the laboratory can be secured in three ways:

1. Much of the chemical equipment must be purchased.
2. Certain items of equipment can be made in the school.

3. Some pieces of equipment can be borrowed or obtained locally as small cost.

The actual stock of equipment and supplies to be kept on hand can be determined only by experience; that on handling in the beginning of the year, should be sufficient for the year.

Chemicals are available in several grades; careful choice should be made since the price difference between grades is large. In very few cases is greater purity than technical grade required for high school chemistry. With the exception of analytical work, there are but few instances when chemically pure or analytical reagent grades are needed.

Semimicro Equipment

The development of facilites for chemistry should take into consideration the use of semimicro techniques. The use of small scale apparatus and very small quantities of chemicals results in considerable savings. The savings in reagents and apparatus, when semimicro techniques are used, are accompanied by a saving of storage space, as well as simplified work areas. In the pvovision of initial installation of chemistry facilities, the investment in laboratory furniture, storage cabinets are fume hoods can be considerably reduced. The small amounts of obnoxious gases produced seldom require the use of a fume hood. There is a practical minimum size of phenomena to impress high school students: many educator prefer to emphasize the use of smaller quantities of chemicals in the ordinay size apparatus.

The use of semimicro technique makes possible the storage of the necessary materials in cabinet shelves between chemistry classes. The materials are then brought quickly to the laboratory tables for the instruction period. Thus the multipurpose science room is readily used for chemistry without the hazard of misuse of chemicals by students in other classes.

The decision to install semimicro equipment and corresponding facilities in a new building must take into consideration the changes in teachers and the resultant problem if a new teacher does not want or does not know how to use this teehnique. A compromise solution is to consider the semimicro technique as a supplement to the ordinary laboratory approach. It can be used to advantage for individual student work of the project type especially where expensive chemicals may be required and in those experiments where obnoxious gases are released.

Miscellaneous Facilities

(1) *Distillation Apparatus.* For the usual laboratory a still having a capacity of one gallon per hour is adequate. The still may be heated by gas or electricity. A gas line or electrical circuit with adequate power must be planed. Such a still is usually automatic and is simple to clean. It is mounted on the wall above a laboratory sink for the water supply and drain. A shelf to support a five-gallon receiving bottle should be provided. An ion-exchange apparatus can be used instead of the still, this apparatus provides water of sufficient purity for all ordinary purposes.

(2) *Waste*, Near the sink or the laboratory tables, stoneware waste jars should be available to receive solid chemical waste. Light-weight containers of plastic or metal may be substituted for the stoneware but they are less durable.

(3) *Key Cabinets.* Key cabinets are desirable in chemistry laboratories when each student has an individual locked drawer. The use of combination locks on drawers make a key cabinet unnecessary.

(4) *Laboratory Truck.* A laboratory truch with acid-resistant shelves is needed to facilitate movement of apparatus and supplies of all kinds. This truk can also be used for distribution of materials for sudent use and so lessen the congestion around the demonstration desk or reagent shelves.

(5) *Reading Centre.* A utility table with chairs in a corner

of he room near the window provides a reading center. If heating and ventilating devices do not interfere, wall shelves under the windows are useful. Reading materials may be displayed on the tables or shelves for student use.

(6) *FireExtinguishers*. Because of the ever-present danger of of fire in the chemistry laboratory, fire extinguishers are necessary within the room. An asbestos or wool blanket should be immediately available to smother buring clothing. A supply of sand is useful in controlling certain types of fire.

(7) *First Aid*. First aid supplies should be housed in a cabinct for safe and clean storage. A chart of first aid directions for accidents peculiar to the chemistry laboratory should be permanently mounted near the first aid cabined. A shower head for a rapid wetting is a common and valuable precaution. All students should be taught to report accidents immediately to the instructor.

FACILITIES FOR STORAGE AND PREPARATION

Furnishing for the Preparation Room

The preparation room serves as a work room for the teacher and for student experimentation and project work. The room should be furnished with a preparation room desk which has all the utilities available in the laboratory. There should be drawer and cupboard space for storage. A separate desk for student work is desirable, and a separate utility table provides work space for a variety of puposes,

At least one analytical balance should be provided for the chemistry laboratory; a good location is on a wall shelf or the balance pier in the preparation room. Such a support is firm and relatively free from vibration.

A work bench for developmental work and maintenance may be provided; in some schools the industrial arts laboratory may be used. This bench should be equipped with hand tools of wood-working and metal-working. It should be mounted on a babinet base to provide storage for tools and supplies.

Stools of adjustable height should be provided for work at the various tables and benches included in this workroom. If a typewriter is available, a speical table and chair are needed.

A filing cabinet for pamphlets and other similar teaching materials can be placed in this room. In some situations the preparation room may serve also as the teacher's office: if this is planned, there should be appropriate furniture.

The Darkroom

Consideration is given to the facilities for the dark-room in Chapter II and in Chapter IX.

Storage and Distribution Facilities

The planning of storage space should reflect the anticipated needs of teaching. For a broad range of student experience with generous provision for individual need, greater resources are essential. Such resources require more storage space and more flexible arrangement because of the lack of uniformity found in them.

Storage for the Teacher. Since the laboratory may be used as a homeroom. consideration should be given to storage of the records and supplies needed for this purpose. The chemistry teacher has many records, including inventory lists. that must be cared for. Books, pamphlets, charts, exhibits and other such materials need to be catalogued and a card file maintained.

Storage and Distribution. Storage of supplies and equipment should be considered in connection with provsions for distribution. The distribution of chemicals and chemical apparatus is more involved than in most other fields. While the problem of dirtribution is generally more difficult in the large school than in the smaller it is an important provision in any school.

Distribution of materials for experimental use by the students is accomplished in various ways:

1. The waterials are placed on the demonstration desk and students take them as necessary under the supervision of the teacher.

2. The materials are kept on open shelves in the laboratory and students use them as required.

3. All materials are dispensed from a supply room by a student assistant.

The methods used for distribution may change during the course of the year. In arranging the fixtures in the laboratory, care should be used in placing them that students do not have to crowd past each other to obtain needed materials. Accessory storage spaces are generally available: the storage of supplies and equipment at strategic points promotes investigation and saves much time. The cabinet base of the sink, fume hood, display case, demonstration desk, and the teacher's desk provide storage space for many items. These facilities may provide sufficient total storage for small school which has included a preparation room and separate storage room in the plans. Additional storage facilities are available in the preparatior room desk, work bench, developing table. wall cabinet and printing table. The student desk should provide for storage of general and individual equipment for the total number of persons using the desk. This kind of storage simplifies the problem of distribution.

Storage Rooms. In larger schools where two or more chemistry laboratories are used, a single storage room serving all of them saves spaces and material. This does, however, require additional facilities for tranporta tion and distribution of supplies and equipment.

While each science room should have separate storage facilities, one storage and one preparation room may serve two science rooms which may be either combination class-room-laboratories or separate laboratories. Both room may be equipped for the teaching of chemistry, or one may be for chemisty and the other for another course. It is essential that storage for chemicals be separate from that for physical apparatus.

A minimum of 125 square feet of floor area should be provided for a storage room. When two laboratories are served by a single storeroom the minimum area becomes 150 feet. If chemistry and physics are combined in the same laboratory, Square the minimum area is also 150 square feet. This is made necessary by stoıage requirements for the physics equipment, most of which is not duplicated in the chemistry equipment., The required area is dependent upon the amount of storage provided in the science room.

The storerooms should be provided with built-in shelving to store the major part of the apparatus and supplies in sufficient amount to seıve the anticipated number of students using the laboratory. Since shelving may extend to the ceillng a ladder should be provided for use in this area.

Built-in-shelving of heights sufficient to hold boules of conventional sizes is advisable. The required amount of shelving can be determined on the basis of the number of students to be served and the quantities of chemicals recommend. A minimum of 1000 lineal inches of shelving is required to store the variety of chemicals needed for one section of twenty-four students. It should be noted that additional sections do not require corresponding additions of storage space. The space recommended is sufficient for materials for three sections; with an increase 50 per cent in size it will serve 6 to 8 sections. Special attention should be given to the ventilation of stored chemicals. Continuous ventilation is very desirable. Such ventilation can be provided as a part of the fume hood installation.

The apparatus storage room should contain shelving carefully designed to house the specific materials to be stored.

This requires a study of sizes of apparatus—particularly if physics apparatus included. Shelf heights of 20 in., 12 in., and 8 in, are desirable. Storage cabinets with adjustable shelf heights have many advantages but the cost is generally greater than that of built-in shelving,

Storage Safeguards. A counter with locking drawers is necessary to store small articles of equipment, A two-or four-drawer index card file provides for the equipment inventory and pamphlet file. A continuous inventory is essential to facilitate the ordering of supplies and new equipment.

Some space, generally a rather limited amount, should be provided in which certain dangerous or valuable chemical and other supplies can be locked. Additional storage space in a different part of the building, such as the basement, is often used to store acids, and other materials which can be purchased more economically in large quantities.

PLANS FOR CHEMISTRY ROOMS

The advisability of a separate room or rooms for chemistry should be studied carefully by those responsible for planning. In many schools provisions for chemistry combined with those for physics. In the larger schools separate rooms are generally provided for the teaching of chemistry. The low percentage of utilization of the chemistry room in the small and moderate-sized school makes the separate room a poor investment. Only the larger schools should consider setting aside a separate room for chemistry facilities; even in such instances, there a good reason to provide for combination classroom-laboratory facilities. Some small but financially able school districts may wish to provide separate rooms for the different sciences.

A Single Room for Chemistry

A single room for the teaching of chemistry through the use of Type F desks. This arrangement is quite satisfactory for a relatively short and wide room. The use of a Type E desk.

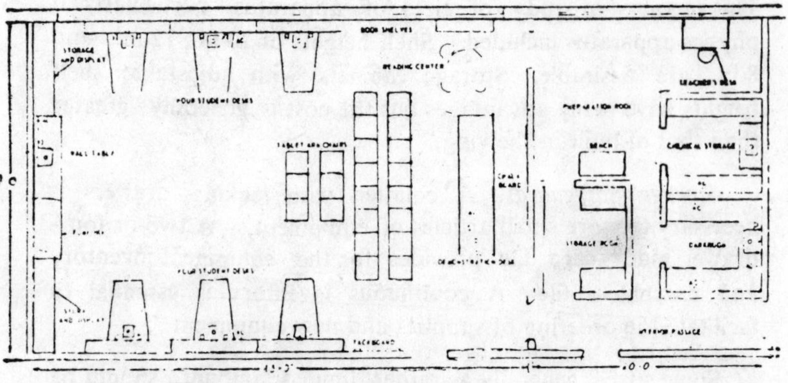

FIG. 5.1. The use of Type H desks along the walls of this classroom-laboratory for chemistry increases the flexibility of the room. The accessory rooms provide for project work and independent student use.

With appropriate provision of services, the room arrangements may be used as a combination classroom-laboratory for chemistry.

A Classroom-Laboratory with Accessory Rooms

The various advantages that result from the provision of accessory rooms are shown in Figure 5.1. There is adequate work space for students in the classroom-laboratory as well as provision for developmental activities. The accessory rooms are available without the interruption of activities which may be in progress in the principal room. The teacher has provision for his own needs, as well as for conferences. It should be noted that the area used for such a combination of classroom-laboratory and accessory rooms does not exceed that used by conventionally planned separate classrooms and laboratories.

Chemistry and Physics with Shared Classroom

A rather conventional plan for chemistry and physics. The shared classroom may or may not have elevated seats. The use of elevated seats tends to formalize the learning situation, makes the room less flexible, and may be ,an accident hazard. However, it may offer some advantage in visibility.

An interesting and useful feature is the combination service window-accessory demonstration desk made available to the classroom by raising the chalkboard, which separates the classroom from the preparation room. The wall table in the chemistry storage room is used in the making of preparations. It is equipped with necessary utilities. Demonstrations can be prepared in advance and transferred to the principal demonstration desk or used on the wall table itself.

Combination Classroom-Laboratory for Chemistry and Physics

The use of 4-student desks in a combination class-room-laboratory. The accessory rooms provide for a variety of needs in both chemistry and physics. The storage cabinets within the classroom-laboratory are recessed into the walls, thus improving the appearance of the room.

The plan Figure 5.2 utilizes fixed laboratory desks so

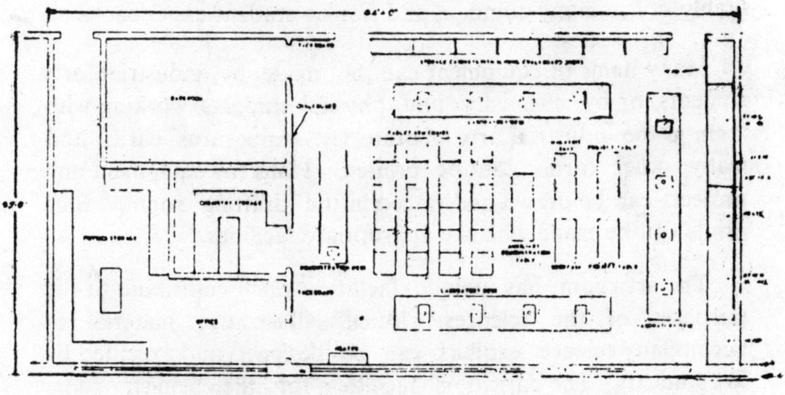

FIG 5.2. This science suite accommodates chemistry, physics, and physical scicnce classes. The work areas are well distributed without lcss of flexibility.

From Munch. Theodore W., and Warren J. Peltcn, "High School Science Facilities." American School Board Journal, 126.55, Januury 1953.)

arranged that a free central area can be used for a variety of science activities. Accessory rooms are used for storage, preparation, and project work.

Alternate Plans

Various alternate plans for classroom-laboratories are provided in Chapter VIII in connection with provisions for teaching chemistry and physics in the same room.

OTHER SCHOOL AND COMMUNITY RESOURCES

Building Resources for Chemistry

It has been suggested that the science rooms be on the first floor near the crafts and home arts rooms, and near the audio-visual center. If the chemistry room is near the industrial arts room, then rooms between them can provide for electronics, forging, welding, and photography. Such rooms should be accessible to both. The tools and equipment of industrial arts should be available to chemistry students. Such fields can profitably utilize many common and similar student experiences.

Many items of equipment can be made by industrial arts students, or by chemistry and physics students working with them in the industrial arts laboratories. Apparatus cards and many other forms can be printed. Plans for equipment and projects can be drawn and traced in the drafting room. Blue prints can be made with the appropriate facilities.

The art room has unique facilities which contribute to the purposes of the sciences. Much illustrative material to accompany science exhibits can be designed and executed by art students. The darkroom facilities for the chemistry room can be used by students in fine and industrial arts.

The students in mathematics can cooperate wilh the chemistry students in such activities as the construction of crystal models and in the use of apparatus to derive mathematical formulas or to demonstrate them.

The stenographic, typing, and duplicating facilities of the business department can be used to prepare material for use in the teaching of chemistry. Duplicating equipment may be a permanent resource of the chemistry room.

The maintenance department of the school, in addition to its ordinary services, can be a source of raw materials such as glass, wood, nails, and wire.

Community Resources for Chemistry Teaching

Each community differs in its resources, but each has some resources for science teaching. The resource which is generally available in communities is the people themselves. Particularly useful are those persons in the community whose vocations and avocations are centered about things scientific. Of particular values in chemistry are the speakers' bureaus of colleges and certain organizations, such as the American Chemical Society. Many industrial organizations provide speakers and consultants.

Direct experience is an essential to effective learning. Fortunate, indeed, is the school located in a community that has a variety of industrial establishments. Factories that make chemicals or products involving chemical processes are effective aids to classroom instruction in chemistry. Industries that produce petroleum products, glass, paints, plastics, synthetic rubber, metals and allows, and many other such materials can be visited by chemistry students.

Food processing plants, ice and cold storage plants, laundries and dry cleaners, power plants, city water systems and sewage disposal systems are other resources of the community with particular importance for the students in chemistry.

Many library resources are available. The public library should be used by students. Industrial libraries may be available for student use and nearby college libraries can usually provide more extensive library facilities than most high school libraries.

Museum and scientific institutes offer displays which can supplement the students' work in school.

Mines and quarries offer an opportunity for chemistry students to see the raw materials of industry taken from the

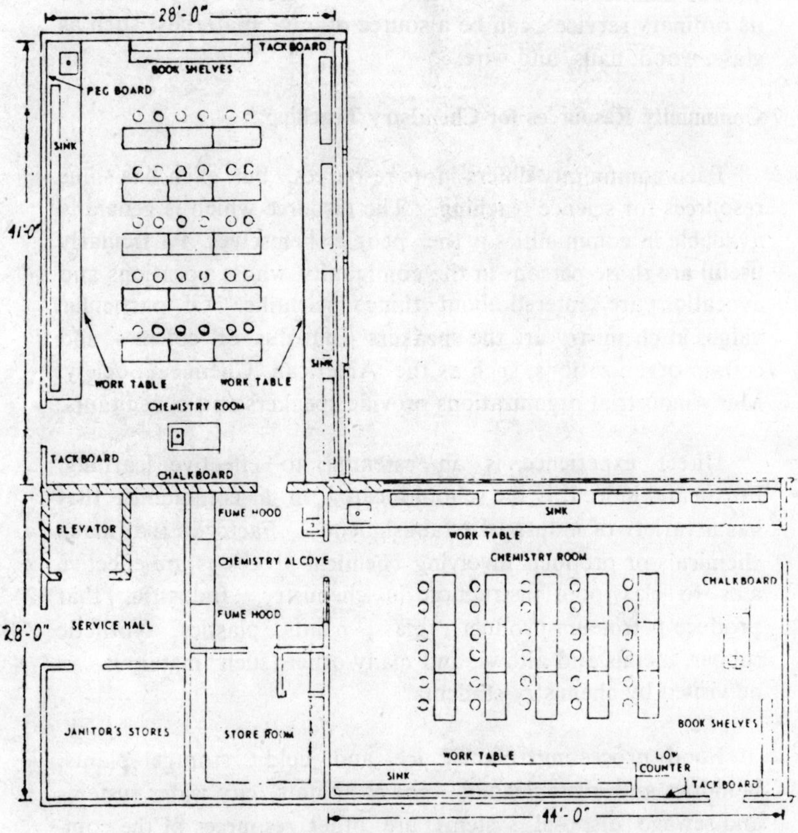

FIG. 5.3. The classroom-laboratories in this chemistry suit eare made flexible by placing the fixed desks with utility lines along the walls. The student tables in each room are movable. Fume hoods and auxiliary work spaces are provided in the chemistry alcove, accessible from both rooms. A small, fire-proof vault in the storage room is vented through the roof. A photographic darkroom is available in another part of the building.
Portland, Oregon, Public Schools

earth. Farms gardens, and orchards provide opportunities for the study of applied chemistry, Greenhouses and nurseries are comparable resources.

Much demonstration material and raw materials for making equipment are available in the community, Scrap materials can usually be obtained locally at low cost. Samples of raw materials and industrial products serve to relate life of the community and the work of the school. With planning, the study of chemistry can become in fact a school-community activity.

CHECK LIST OF CHEMISTRY FACILITIES

1. Do the facilities provide for a broad range of experimental work in teaching chemistry?
2. Are there adequate facilities for using demonstrations in teaching chemistry?
3. Do the facilities for chemistry make possible construction and other essential developmental activities?
4. Are resources for display and illustration available to facilitate the teaching of chemistry?
5. What projection equipment is provided?
6. Are facilities for reading and writing provdied?
7. Do the maintenance and safety provisions meet the unique needs of chemistry teaching?
8. If a single room only is to be used for chemistry does it make provision for storage, preparation, photography, and other activities usually found in accessory rooms?
9. Can an accessory room or rooms be provided?
10. Does the major room for chemistry provide for both normal classroom and laboratory activities, or are separate rooms to be provided?
11. If the classroom-laboratory for chemistry is to be

shared by other science (or non-science) classes are there adequate provisions for the needs of such classes?

12. Does the arrangement of the furniture and furnishings in the chemistry room provide for a maximum flexibility in their use?
13. Do the provisions for floors, decoration, and ventilation take into account the unique needs of the work with chemical reagents?
14. Are the provisions for the control of obnoxious or dangerous gases adequate?
15. Are the utilities anticipated adequate to a broad range of student experimentation?
16. Does the distribution of utilities in the room or rooms promote the most effective work in chemistry?
17. Can the necessary facilities for the semimicro method of chemistry work be installed profitably?
18. Have the essential miscellaneous facilities been anticipated?
19. Does the preparation room provide for such activities as chemical preparations. construction work, storage, and project work?
20. Do the facilities provide for photographic work?
21. Are the facilities for storage and distribution of reagents and apparatus adequate?
22. What building resources can be used in the teaching of chemistry?
23. What community resources can be used in the teaching of chemistry?

BIBLIOGRAPHY

1. American Association of School Administrators, American School Buildings. Twenty-Seventh Yearbook, Washington, D.C.; The Association, 1949, pp. 93-97.

2. Byerley, J. Roy. "Planning and Equipping the Science Laboratory," American School Board Journal 98: 59-62, January, 1939.

3. Campsen, Herman M., Jr. "Purchase, Construction, and Use of Science Laboratory Apparatus," The American School and University, Seventeenth Edition. pp. 382-386, 1945.

4. Engelhardt, N.L., Sr., N.L. Engelhardt, Jr. and Stanton Leggett. Planning Secondary School Buildings, New York, Reinhold Publishing Corporation, 1949.

5. Committee on Science Education in American Schools, Science Education in American Schools. Forty-Sixth Yearbook, Part I, of the National Society for the Study of Education. University of Chicago Press, 1947.

6. National Council on Schoolhouse Construction, Plant Guide Committee, Guide for Planning School Plants, Nashville: The Council, pp. 58-60, 1949.

7. Herrick, John H., "The Development of Educational Specifications in Planning the Secondary-School Plant." Bulletin of the National Association of Secondary School Principals, 32: 213-219, March. 1948.

8. "School Plant Trends," The School Executive 68:36-91, January, 1949.

9. Williams, Harry H., "Moving to Semimicro in High School Chemistry," The Science Teacher 17: 172-174, November, 1950.

6

School Science Facilities

INTRODUCTION

The basic principles discussed in privious Chapter that should serve as guides to the development of facilities are grouped into the following discrete categories:

General Planning of Science Facilities

Characteristics of Rooms for Science Teaching

Facilities for Student Work in Science

Audio-Visual Facilities for Science Teaching

Facilities for the Science Teacher

The implementation of various principles within each of these categories reveals ideas, procedures, and physical requirements concerning science facilities that are of general significance; such aspects are related to the teaching of science at all levels and in all fields. In other ways, many of the principles have specific and unique meanings for different levels and different fields. The aspects that are of general significance are considered in the present chapter under the appropriate categories; those that have specific and unique meaning for different levels and different fields are included in the appropriate chapters.

GENERAL PLANNING OF SCIENCE FACILITIES

Selection of Site

The first step in the realization of plans is the choice of building site. In those situations where remodelling is planed, this choice may not exist, or it may be a choice as to which school in a system having several buildings is to be remodeled. In other instances where a site is to be chosen many factors enter into the choice. Among such factors are the availability of the land, its cost, the geographic dietribution of the students, trends in home building, and the like. The nearby potentialitics for science teaching should also be a factor, weighed along with other factors in making the choice of site. The availability of wooded areas, streams, lakes, or marshes; experimental farms, orchards, and gardens; natural formations; factories, quarries, research laboratories, and other industrial and technological developments, and similar resources should be considered. In those communities where a long-term policy of land purchase for school use is practiced, the needs of science teaching and the potentialities of anticipated building sites should be carefully weighed, just as they should be where building is immediately planned.

One of the functions of the school architect is to landscape the school grounds. A board of education sensitive to the potential values in the nearby surroundings can encourge the architect to preserve those features which can enrich the science work. The utilization of such facilities is largely a function of the teachers of the school; use is not possible unless these resources are provided by those responsible for the general development of school plans and policies.

Background for Planning Science Facilities

The expansion, replacement, and improvement of school facilities require careful planning. Buildings are obsolescent, school population changes, and understandings of the charactetistics of effective facilities grow.

Effective planning involves many persons, Ideas, information, and points of view gathered from various sources can be synthesized by competent leadership and translated into action which results in desirable school buildings and adequate learning facilities. Not the least of the values derived from cooperative planning is the support gained from those who participate.

It is necessary to plan facilities so that the learners may obtain the useful contributions of the biological sciences, the physical sciences, and the earth sciences. Many educational leaders insist that this will require first a planned program for science in the elementary school. To this, they insist, must be coupled a planned science programme for grades seven, eight, and nine, and for the senior high school. Such an integration has long been recognized as a desirable goal, but is seldom achieved. The recommendations of these leaders would include for the senior high school at least a full year in the biological sciences and a full year in the physical sciences, including some earth science. Beyond this they favour opportunities for students to elect further work in the biological, physical and earth sciences as their interests and ambitions may lead. and opportunities in the community and world may require. Some educational leaders argue that twelve or fourteen years of science studies are necessary in an age teeming with scientific devices and associated social problems.

A major problem is, then, to determine what science experiences and programmes are to be a required part of the education of all youth in the community and what experiences and programmes are to be open to electron by students.

The acceptance of this point of view is sometimes described as "planning from the inside out." The facilities for learning in science should be built around the needs of young people, as those needs are determined by the society of which they are a part, Choices of materials, apparatus, furniture, rooms, buildings, and social sites are based on the kinds of experience necessary and desirable. The bases for the choices and decisions should be understood by all concerned.

Personnel of the Planning Committee

The planning of science facilities is the joint responsibility of several qualified persons: no one person should be expected to be capable of performing this entire task. The unique competencies of the administrator, the science teacher, the specialist in the teaching of science, the school building consultant, and the architect should be brought to bear upon the task. There should be representation of the community, perhaps through the Board of Education or through the Parent-Teacher Association, or both. There may profitably be representatives of students in school, and of the alumni, of teachers from related fields, and of the custodial staff of the school.

A committee with representation of the teachers, administration, school plant specialists or consultants, architects, the community, and the students should be organized to plan the science facilities. The groups represented and the number of representatives from each group is a matter of local decision. The science group may be a subcommittee of a larger committee responsible for the planning of all school facilities.

In schools of small and moderate size all teachers who are directly concerned should be members of the committee. In planning the science facilities for an elementary school, for example, all elementary teachers, as well as the science teacher or consultant, should be involved. In secondary school of small and moderate size, all science teachers should be involved. In large school systems, it is more desirable ta provide for representation on such a planning committee. In those systems that have supevisors (elementary or secondary) they, too, should be members of the committee.

Those teaching fields most closely related to science should be represented. Teachers in such areas as industrial arts, home economics, mathematics, and agriculture can make significant contributions to the planning.

Students and alumni are in strategic positions to assist in the planning of science facilities. It is they who have been and are direct beneficiaries of the science resources of the school. From them and the gropus they represent can be drawn ideas and insights as to needs; from them much support can be gained in the present and future years.

To assist the science teachers and others concerned with science facilities, a leader in the field of science education should be made a member of the planning committee or invited to review the plants while they are being mede and before they are carried out.

Education is basically a function of the community. Representation of parents and others in the community is highly desirable on the committee responsible for planning science facilities. They can assist greatly in determining present needs and future trends in community growth.

Functions of the Planning Committee

A general statement of the functions of the responsible committee has been set forth. According to this statement the committee should:

1. Evaluate the existing school purposes and programmes.
2. Recommended changes in the existing purposes] and programme of the school.
3. Prepare a statement of the educational and building requirements for the school.
4. Assist in the preparation of reports on buiiding requirements.
5. Assist in the evaluation of basic building plans and specifications.

6. Assist in the preparation of equipment layouts and equipment specifications.
7. Assist in the evaluation of completed buildings.[13]

Various specific activities can be undertaken by the science committee as a whole or through subcommittees. Such activities include.

1. A study of the needs of the community that can be met through an appropriate science programme in the school.
2. A study of the science enrollments in the various courses in science over a period of several years, and a determination of trends and probable future needs.
3. A study and projection of shifting enrollments which would probably be brought about by a changed emphasis in programmes, and by the presence of different facilities.
4. A study of the pertinent literature, Such literature may include curriculum studies, community surveys, school building studies and proposals, and the like.
5. Determination of the space requirements, both in classrooms and in accessory rooms.
6. A pooling and discussion of ideas and proposals, resulting in recommendations and plans.

The science teachers should be leaders and active participants in such activities. It is they who should know the function of the various fields of science in the life of the community and in the school programme. It is they who will use the facilities as they are realized. The work in science can be no better than the teachers who are responsible for it. The quality of their direction must become evident in the planning stage. Whereas various committee members can do little more than assist in determining policy, the science teachers should be not only leaders in determining policy, they must also supply the details concerning supplies, equipment, and building facilities.

The committee on science facilities should bear in mind that science is only one phase of the curriculum and that science should be interrelated with other aspects of school work. Other committees are planning for other areas of activity. It is essential that committes work closely togather, not only to integrate policy concerning the over-all planning, but also to give attention to details of planning that are of general concern.

A Word of Caution

It is impossible to set up the specifications for an ideal science learning situation without reference to a particular school. The need to identify science instruction with home and community living requires local planning. While the ideas and suggestions presented in this volume can not serve as a compresensive list from which to select and adapt, they can provide a stimulus to groups concerned with planning if it is rembered that the facilities planned must be related to the immediate situation.

Few school systems can afford to indulge their every with in the matter of desirable science facilities. But even if unlimited resources were available it would be impossible to design a perfect science room. The importance of the individual in a changing school population and the never-ceasing evolution of the curriculum are factors which preclude this possibility Groups concerned with planning science facilities should realize that needs cannot remain fixed. They must, therefore, be judicious in selecting adaptable equipment, flexible furnishings, and generally useful supplies.

Use of Planning Kit

Space planning can be a very time-consuming time-consuming task. When completed by the use of the usual local resources only it is commonly vague and unrealistic. To facilitate effective planning of science rooms, kits of models that represents rooms and major furnings for science teaching are available. In such kits the floor plan is flexible in size and design, and the tables, cabinets, cases, and similar items are prepared on the same scale as the floor plans. Thus it is possible to develop a variety of floor plans that are suited to the needs of the community. These rooms can then be arranged as desired and the furnishings measured. Such plans provide effective communication among those responsible for planning, those who administer the plans, and those who will use the science facilities.

Use of a Check List

The committee on science facilities should be concerned at every step of the planning process that the facilities as planned will serve the principles set forth in Chapter 1. While it would be impossible to present a check list that would enable a committee to verify each detail in every situation, the principles themselves become criteria against which details are checked. A general check list is provided on page 49.

CHARACTERISTICS OF ROOMS FOR SCIENCE TEACHING

Rooms used for science teaching have many common qualities, regardless of the level or the field of science. Space requirements, illumination and heating, attractiveness, and provisions for flexibility are among the common characteristics to be considered in planning.

The Space Required for Science Teaching

The determination of needed space within the building of necessity based upon the quality of the anticipated programme. Proposals for the provision of such space are set forth in the various chapters that follow. Essential to the allocation of space is the fundamental question of the number of rooms to be provided.

The calculation of the number of rooms required (classrooms or laboratories, or combined classroom-laboratories) consists of determining the total number of student-periods of instruction to be provided and dividing this number by the number of student-periods of instruction that can be provided by one room.[8]

The formula is:

$$\text{Number of rooms} = \frac{\text{student-periods}}{\text{average class size} \times \text{number of periods per week}}$$

The quantity, "student-periods." is determined by multiply ing the number of students in a given course by the number of times thas a given class meets per week. For example, 275 students in general science meeting five times per week would require 1.57 rooms, assuming a school day of 7 periods and an average class size of 25.

$$\text{Number of rooms} = \frac{1375}{25 \times 35}$$

$$= 1.57$$

Calculations for other rooms are made accordingly. If the sum of all major rooms needed is less than one. then a multi-purpose classroom-laboratory should be considered. If the sum is nearly two, appropriate combinations of rooms should be sought for maximum utilization.

In many instances it is necessary to use a science room for a course outside the field of science. If this is planned, the other course to be accommodated should be studies and the calculations made in the same way.

Area Raquirements per Student

The teaching of science in modern schools requires a larger area for each student than was provided in earlier facilities. The bases for this increased area are:

1. The nature of learning activities, particularly in science,

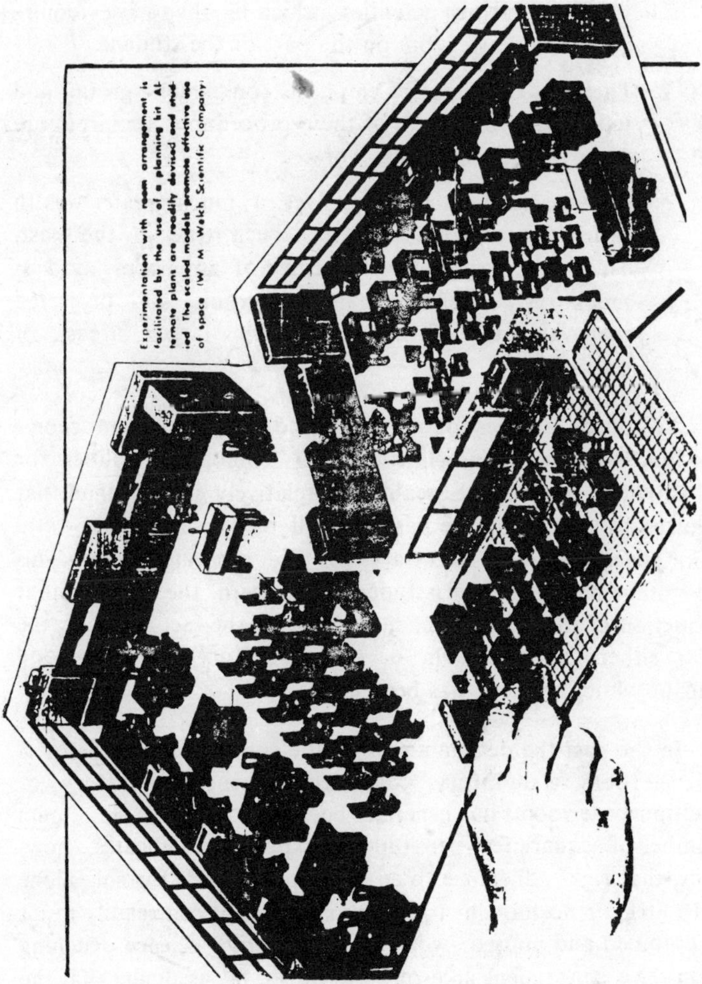

Experimentation with room arrangement is facilitated by the use of a planning kit. Alternate plans are readily devised and studied. The scaled models promote effective use of space (W. M. Welch Scientific Company)

has been broadened to include a greater range of activities than in the past. Within this increased scope there are various activities which in themselve require considerable freedom on the part of the student.

2. There is a greater emphasis on small group and individual work; each of these procedures require more space than formal class work.
3. Science teaching now involves a much greater wealth of learning materials than has been true in the past. There is a greatly extended array of apparatus, models, mock-ups, as well as charts. diagrams, and projected materials, all of which require larger areas of operation.

As science facilities follow the trend from separate rooms for laboratory work and class work to combined facilities for all activities, per-student area can be relatively greater than that allotted to students in the past and still require a smaller total floor space than was necessary for the separate rooms. This advantage of a larger per-student area with the concomitant reduction of total area is made possible by the fact that in the past either the classroom or the laboratory has often stood vacant while the other has been in use.

In the past the determination of the needed floor area of a science room (laboratory, combined classroom-laboratory or multipurpose room) has generally been based upon a minimum number of sequare feet per student. Experience indicates, however, that rigid adherence to an area-per-student formula alone with little or no thought to room arrangement can readily result in cramped and unused work areas. Effective science teaching requires a per-student area of 35 to 40 sq, ft., assuming that the room arrangement makes possible the utilization of all available space and facilities with a minimum of difficulty. This calculation does not include such accessory space as storage rooms, preparation rooms, and darkrooms. If such rooms are included, the per-student area should be at least 50 sq. ft.

Efforts have been made to determine the relation of storeroom space to the total area for science facilities. Certain estimate indicate that the storeroom should utilize approximately 20 per cent of the total area. This figure can be only approximate since it is dependent in part on the provision of storage within the classroom-laboratory itself. It depends also upon whether the classroom-laboratory has multiple functions. If so, the storage facilities should be correspondingly greater. There is adequate justification for providing much storage space within the classroom-laboratory itself, in which case a relatively smaller separate storeroom should be adequate.

Specialized Science Laboratories

Experience indicates that high schools of fewer than 500 students should provide for multiple use of science rooms, so that as a minimum one classroom-laboratory space serves both general science and biology, and another serves chemistry, physics, and other related courses. In the smaller high schools it is advisable to provide a single multiple-purpose classroom-laboratory for all science teaching.

In schools larger than approximately 750 students, there may be justification for separate rooms for chemistry, physics and biology. The decision on separate rooms depends. In part, upon the nature of the science programme in the school and upon the number of students electing courses in science. In such instances the facilities may well be grouped on one floor in one wing of the building, the orientation following the principles set forth.

Large schools may profitably consider a separate building for science or for science and such related areas as home economics and industrial arts. There are distinct advantages in plan: there is greater flexibility in design, and there can be better grouping of facilities. The disadvantages in a separate building include a tendency to further separate science from other aspects of the school curriculum, and in cold climates there are increased problems of heating.

Provision should be made for such common needs as storage, work space for the teachers, preparation of demonstrations, and construction-type activities. It is desirable to limit the storage in any one storeroom to two different areas, such as general science and biology. Providing common storage for chemistry and physics introduces certain problems. The corrosive effects of the vapours from certain chemicals on physics apparatus is undesirable. White the problem is serious, it can be solved if three precautions are taken:

1. the physics apparatus subject to damage should be stored in cabinets with doors;
2. the storage room as a whole should be well ventilated; and
3. continuous ventilation should be provided for the stored chemicals.

Whatever the layout of the laboratories for specialized science map be, it is desirable to provide a separate space for preparation and construction-type activities. The provision in such areas depend upon the teaching needs to be served.

The Design Requirement and General Features

The requirements of floors, walls, illumination, heating and ventilating, plumbing, and electrical service vary from one school situation to another. Many of the requirements are such that they can be reduced to general statements which, with some modification, are appropriate to most schools.

1. *Floors.* The essential factors to be considered are comfort, resistance to chemicals, safety, and attractiveness.

Comfort, is especially important in view of the fact that much of the activity in science depends on standing and walking.

The floors of rooms used for chemistry should be acid-resistant. The use of chemicals in other science courses is so limited that the floors need not be acid-resistnnt. If an extensive use of liquid petroleum products is anticipated the floor should not be covered with common asphalt tile or mastic composition because of the deterioration caused by such liquids.

The material used in floors should be free from unnecessary polish in the interest of safety and freedom from glare. The floor surface when wet should be relatively safe from danger of slipping.

The colour of the floor should be relatively light, in the interest of room attractiveness and good visual environment. Certain of the vinyl plastics give promise of providing the necessary and desirable qualities of floors.

2. *Walls and Ceilings.* It is essential that care be exercised in in the choice of materials for the walls and ceilings of science rooms. In certain kinds of situations, such as the use of sound motion pictures, the radio, and recording equipment, the acoustic qualities of the walls and ceiling become important. Because of the fire hazard in the science room, the acoustic materials chosen should be fire-proof.

A second quality to consider is the paintability of the acoustic material. Some acoustic tile loses its acoustic qualities when painted. The science room may need redecoration more often than other rooms; the original choice of acoustic material should provide for this need.

3. *Illumination.* The principal unique needs in the illumination of the science rooms have to do with the provision of extra light in spaces where tasks requiring close scrutiny are to be performed, and where students may be somewhat removed from objects to be observed.

Those areas requiring extra light because of the exacting work to be done there include; balance shelf or pier, microscope shelf or table, tables where electrical or other instruments with fine scales are used, and the like. If the areas are permanently set aside for such use. permanent fixtures for auxiliary lighting should be installed. For greater flexibility it is probably more desirable to use auxiliary lamps that can be plugged in as they are needed at convenient points.

Those spaces that are often scen at a distance include the demonsiration desk, chalkboards, tackboards and exhibit cases. In general, the provision for auxiliary illumination for such spaces should be made permanent. The circuit for each should include a switch for independent control. It is sometimes desirable to focus attention on the demonstration desk by darkening the remainder of the room. Directional lighting units placed in the ceiling above the demonstration desk are used in some schools. Auxiliary lamps can be used on the demonstration desk. To provide for this there should be an extra duples electrical receptacle on the demonstration desk.

4. *Heating and Ventilation.* The special needs for heating and ventilating science rooms are principally of the following kinds:

(a) Adequate and independent ventilation should be provided for animal rooms at all times. In certain respects it is better to separate the animal room from other classroom and workshop facilities. With adequate ventilation the odour of such a room does not become obnoxious. Such provisions reduce the problem of the allergic rcaction of some persons to fur and feathers. The provisions for temperature and ventilation control in the animal room should provide conditions as nearly like those of the natural habitat of the animals as possible.

(b) Areas where obnoxious or dangerous vapours may be produced, such as the chemistry laboratory, should have vents with exhaust fans directly to the out-of-doors. Such vents should be provided both for the demonstration desk and for the student work areas. It is inadvisable to attempt to use a common duct for normal exhaust from the building for the discharge of obnoxious chemical vapours. The duct which carries chemical vapours should be coated with a chemicaily-resistant plant (such as a paint with an asphalt base) or be constructed of glazed tile.

(c) Continuous ventilation should be provided for chemicals that are in storage. Certain of these produce vapours that are corrosive and in some iustances obnoxious and even dangerous. Such vapours should be continuously removed by proper venting. It has been found possible to provide for the storage of chemicals needing such venting within cabinets connected to a fume vent of the quality described in (b).

(d) It is particularly desirable to provide adeqvately for areas that have little or no outside ventilation. Such spaces as the storage and preparation room, the photographic darkroom, and the planetarium (if this is included) should be given special attention. The use of such rooms by several persons at one time without adequate ventilation usually leads to undesirable conditions.

(e) Particular attention should be given to the provision of constant temperature for plant growing rooms. Provisions for the control of the temperature in such rooms should be independent from the controls for the building. The reduction of over-all temperature at night and during week ends and vacations should not require that growing rooms and animal rooms also be rendered detrimental to life. Heating coils buried in the soil of terraria or plant beds provide the desired control if they are thermostatically controlled. An auxiliary atmospheric heater can be similarly controlled.

5. *Plumbing.* The plumbing requirements of science rooms are of the following kinds:

(a) The extent of the provision of running water, in rooms used for science teaching depends upon several factors, with perhaps the major one being tne specific purpose or purposes for which the room is to be used. Running water has so many uses in the teaching of science that there seems to be no exception to some provision for its use in every room used for science teachin. Wherever possible there should be both hot and cold running water. Such a source is a means of saving time in many laboratory operations, and facilities the cleaning of glassware and other operations.

(b) The materials used for sinks, troughs, and drains must be selected in the light of the demands placed upon them. They must withstand the corrosivc effects of acid and bases. in particular. Other chemicals, such as strong oxidizing reagents, sometimes place special demands on waste lines. Mercury, which is often used in the science laboratory, [amalgamates with lead and causes deterioration.

Several different materials are available for use in sinks and troughs. One of the more commonly used materials is soapstone. Another is impregnated sand-stone. Sinks and troughs made of slabs of these materials carefully joined and cemented are quite satisfactory. Ceramic ware sinks have the advantage of being made in one piece. The use of sinks mape of procelain on iron or steel is quite satisfactory of certain kinds of situation. It has the disadvantage of being readily chipped by falling objects, From the standpoint of chemical activity, lead-lined sinks are the most inert. They have the disadvantage of being damaged chemi-cally by mercury which amalgamates with the lead. Other materials which are satisfactory for ordinary use in laboratory plumbing are stainless steel, Monel, Duriron, and high silicon cast irons.

The material used in waste lines should be chosen with care because of the specialized demands placed upon it. In general, high silicon iron, chemical stone-ware, cast iron. and lead are satisfactory.

(c) Sources of heat for the various experiments and demonstrations and for laboratory operations are essential in rooms used for science teaching. Probably the most satisfactory source is gas. Outlets for gas lines should be provided at several points in each room; the details are suggested in the appropriate chapters. In communities without natural or, locally produced gas, gaseous fuel liquefied under pressure is often used. For many types of operation, alcohol burners and electric hot plates are satisfactory. Such sources of heat thould be available in all science laboratories, regardless of gas supply, for use during emergencies, peak demands on sources of heat, and for the occasional need of a source of heat in a room not supplied with gas.

(d) Some school buildings are equipped with an air compressor. If a source of compressed air is available, it should be piped to the science room or rooms for use with blast lamps, in experiments concerned with air pressure, and the like. If there is not a general source of compressed air available, a portable unit should be provided for use in the teaching of science. Any air compressor with a storage tank should be equipped with an automatic control that shuts off the compressor when the pressure reaches 35 lb, per square inch and turns it on when the pressure is reduced to 25 lbs per square inch.

(e) For water, gas, and compressed air lines, there should be a master shut-off valve for each science room. Such a valve facilitates safety practiee and repair work.

6. *Electricity.* The provision of electric service to the various room is dependent upon the anticipated function of each room. The use of various pieces of apparatus that depend upon 110-v. a.c, is such that every room used for science teaching should have several such outlets. The outlets should be well distributed. The particularized needs of each room are considered in the separate chapters.

7. *Clocks.* In the usual school a system of clocks for the entire building is installed. Such a system utilizes a master clock with which clocks located at varions points in the building are synchronized. A clock in the time system should be located in each of the rooms used for science teaching, including the accessory rooms. Such clocks are essential to laboratory and other informal learning activities where the responsibility for the use of time rests in part upon the students.

Attractiveness of Rooms

The esthetic qualities of science facilities have proably been among the most neglected of all the physical factors. In spite of the fact that highly polished furniture and gleaming equipment in a new laboratory make an attractive picture to some science teachers, we must recognize the potential danger in the glare caused by the reflection of light. The esthetic appeal of a diagy room with its dark floors. black table tops, dirty walls. and unpleasant odours is negative to all but the least sensitive of persons. Science rooms should be attractive,

1. *Decoration*, There are certain basic conditions of appearance upon which probably most teachers can agree. Among these are adequate illumination, the general use of light colours rather than dark for walls and furniture, the use of clean-cut lines in the plans, avoidance of exposed, unsightly structures, cleanliness, and order and neatness.

Colour preferences are related to room exposure by many persons. There is rather wide agreement on the use of the so-called "warm" colours for the walls of the rooms having a northern exposure. Colours and tints such as ivory, peach, and buff are often used: yellow is favoured by many for sunless rooms. Light shades of green are often preferred for rooms having a southern exposure.

The attractiveness of the walls of a rcom is related to their reflectance of light and to the durability of the paint. A high gloss should be avoided in the interest of eye-comfort as well as attractiveness, Paints of high durability from the standpoint of resistance to corrison as well as to the process of washing are those having a rubber base, those of vinyl and acrylic types, and enamels with an alkyd or urea alkyd base.

(2) *Design.* In the planning of new buildings and in reconditioning those that are old. it is possible to improve the attractiveness of rooms and furnishings by giving attention to regularity of structure. The avoidance of miscellaneous type and sizes of furniture and the giving of attention to balance in the arrangement of fixtures and accessories on wall areas are examples of steps that can be taken in this direction.

(3) *Furniture.* The selection of the furniture should take itno account the reflectance of the finish. Glossy surfaces should be avoided; the stain, if any, should be relatively light to provide for a more favourable reflection factor. The elimination of the black top of the laboratory table improves the appearance of the room as well as the quality of the illumination. For some laboratory operations the acidproof table is essential; recently developed plastic coatings for wood are proving to be quite satisfactory in meeting the serve demands placed on table tops in the science laboratory. Such plastic coatings are being produced in attractive colours, For many purposes, a well-varnished surface (dull finish) with careful maintenance is adequate. A waxed linoleum surface is adequate for most laboratory demands; such material in some light colour is very attractive, Other materials that have been found satisfactory for many operations are cement asbestos, Masonite and Formica, These materials are somewhat more attractive than the familiar black-top.

(4) *Fixtures and Accessories.* Attention to such fixtures of the science room as the chalkboard and tackboard can improve its appearance materially. The familiar "blackboard" absorbs light, often to the detriment of illumination and proper brightness balance. Chalkboard made of glass in light shades of green and yellow do much to make the room attractive because of the colour and the increased reflectance. Likewise tackboards

should be obtained in light sbade; most tackboard materials can be painted.

The science teacher can assist in the improvement of the appearance of the facilities through the use of decorations. Appropriate pictures in bright colours can be used. Good murals often add to the atmosphere of the room; in many instances they can be produced by the students. Continuous attention to the displays on the tackboards is essential. Colourful materials that are carefully arranged and periodically changed do much to improve the appearance of the room, Black and white pictures that are to be displayed on the tack-board can be mounted on a backing of heavy coloured paper for better appearance. If carefully maintained, growing plants, aquaria, and terraria can add to the attractiveness of science room.

Flexibility of Science Facilities

A major consideration in the planning of facilities is the fact that the future of science teaching is difficult to foresee; this, in itself, indicates that the teaching facilities should be flexible, The curriculum is continuously emerging and changing. The educational service rendered by a school plant is of changing character, just as the very nature of communities is subject to the change.

The school curriculum should reflect the evolving need whieh it is to serve. Such emphases as cooperetive technique in planning and life experience as a basis for curriculum find their expression in school facilities as "open plans" with flexible layouts and movable furnituse, and the planning of classroom for ready alteration.

Educators have found that the maximum use of school resources results from the kind of planning that makes varied uses of facilities possible. The provision of specialized facilities tends to limit that use. This is often true of science facilities. Because of the cost of such resources for learning, every effort should be made to provide conditions for optimum desirable learning. Careful selection of furniture and equipment and provisions for varied activities in a given room make possible more and richer experience than is possible in rooms made inflexible by highly specialized equipment.

In the elementary school the matter of achieving flexibility is not so difficult as in the secondary school. The major fixed installations needed for elementary science are a source of running water with drain, gas, and electricity. Within the limitations of such fixed facilities the necessary table space for the work of elementary pupils and the storage space needed for supplies and apparatus can be adjusted to the needs of the situation. In general, it is desirable to place the three given facilities near each other, although they should not be so close that their use procedures crowding. Thus it is possible to provide a science centre within each classroom. If a separete room is to be used for science in an elementary school, it is desirable to have two or more such centers for service connections in the room. The use of movable furniture in either the classroom so separate science room helps to achieve the flexibility or desirable in science work.

In the construction of the small and moderate-sized high school it is desirable to plan flexible facilities because of the demonstrated need for multiple use of facilities and because of the changing pattern of need within a school over a relatively few years of time. Experience indicates that science rooms and their facilities in such sehools tend either to stand unused or are used by classes in other fields, thereby limiting student work in science. The plan for facilities should be such that students in science can work in a separate room with science facilities while the major room is used during certain periods by students in another subject. The separate room can be a project room, preparation room, plant and animal room, dark room or. perhaps, a storage room.

In the large high school needs often change over a period of years, particularly as enrollments in elective courses tend to change. Such patterns shift as a result of the interests of teachers, developments in the field of science which affect the curiculum, and nceds of the times such as those occasioned by national emergencies. There is also the need in both large and small schools for providing certain special facialities for talented youth whose needs require more than coaventional facilities. Ways of achieving flexibility have been rather well summarized in a bulletin issued by the Department of Education of the State of Alabama:[9]

1. Make internal partitions non-bearing where possible.
2. Windows should be continuous, not grouped.
3. Place conduits, vents, water pipes, etc., in corridor and outside wails.
4. Group rooms in series, according to function.
5. Design cabinets, lockers, and shelves as portable units.
6. Use movable partitions and folding doors where possible.

The planning of new buildings should envision a use of modular construction as well. A module is . . . "the smallest repetitive unit of space and must be complete in its repettiion of the characteristics that enclose and serve this space. The characteristics of this repetitive element are its three dimensions, its architectural, mechanical, electrical and structural features and the services that may be added for the convenience of its occupant."[10] The opportunity to shift partitions in relation to the unit of structure of the building makes possible an easy adjustment in room area and proportions. Folding doors well insulted against the transmission of sound make further adjustments in room size and use, Within the room itself the use of storage cabinets and shelves makes possible the division of the room into separated work areas without the inflexibility whicn results from fixed walls and partitions.

Flexibility Through Multiple Use of Facilities

The use of seprrate classroom, or lecture room and adjoining laboratory, leads to inflexibility and to a lack of use of science facilities during a major portion of the school day. The use of a classroom-laboratory avoids this problem and makes possible the ready transition from one kind of activity to another. It also relieves the inconvenience and inflexibility involved when a classroom is shared by a seience teacher and a teacher from another field.

The National Council on Schoolhouse Construction recognizes four distinct types of plan for science rooms:

1. Separate rooms for demonstration and lecture, and for experimentation and laboratory work,
2. A laboratory with one area having seats and demonstration desk, and another area equipped for student experimentation.
3. A laboratory equipped with one-way facing tables, the entire room to be used for all science classes and activities.
4. A laboratory for all science activities, with fixed tables around the walls, a demonstration table and chairs in the front and movable tables near the rear of the room.

This guide points out that type 3 and 4 conserve space, and make programming easier and more flexible.[12]

A study was made by Hart and Givens to determine the reactions of teachers to combined classroom-laboratory facilities. The teachers who preferred comblned facilities gave the following reasons for their preference:

1. There is flexibility in the use of resources.
2. The combination of teaching methods is easy because resources are near.
3. Methods of teaching do not separate themselves by physical pattern.

4. The organization of the class is facilitated; for example, small groups may be easily set up.
5. Psychological aspects of problem-solving are promoted.
6. Students tend to learn by wholes.[13]

Potential Expansion

A study of the community may reveal that a sharp increase in facilities for science teaching will be necessary within a period of a few years. It may be inadvisable to make the necessary installations at the time of building. The task of conversion can be simplified and made less expensive by relatively simple measures:

(a) The provision of one or two (or more) general classroom in an area adjacent to the science rooms being planned. Such rooms should be capable of conversion.

(b) The avoidance of load-bearing partitions at the ends of science rooms; movable partitions make enlargement relatively easy.

(c) The roughing in of water, sewer, and gas lines to points of anticipated need, and the placing of electrical outlets at desired points.

(d) The avoidance of plumbing and electrical conduits so far as possible in removable partitions.

(e) Placing the science rooms in a wing of the building so constructed that additional rooms for science teaching can be added to the wing by new construction.

Orientation of Facilities

(1) *Exposure.* The provision of direct sunlight for plant growth in general science, biology and multipurpose science rooms requires south-facing windows in the rooms for these courses. The growth of plants is possible, of course, in rooms limited to eastern or western exposure, but southern exposure is considered essential by many teachers. The science teachers

should keep in mind, along with the need for southern exposure, that it is important to maintain a low brightness-difference ratio in the illumination of the science room. The provision of this general quality of the illumination of all rooms falls within the purview of the school architect.

(2) *Floor Location,* A major problem of the location of science rooms is that of the choice of floor in a multi-storied building. As with certain other problems, there is a conflict of values involved in this decision. There are advantages in the use of the first floor for the science facilities. Chief among these advantages are the ease of use of growing areas in or near the school building, the simplicity of leaving the building for field trips to other near destinations, and the relative case of bringing into the laboratory materials from the outside. The latter is particularly important in the case of large and heavy objects. There is likewise a certain advantage in the reduced cost of the specialised plumbing to the first floor rather than to a higher floor. For rural and suburban communities the first floor location is probably more advantageous than it is in the large cities.

There are some advantages in the use of an upper floor. In the planning for certain phases of science it is desirable to have access to roof areas. Whether and astronomical facilities may be located there as well as a growing room and a redio workshop and transmitter. Access to such facilities can often be provided directly from the science suite if the suite is located on the top floor of the building. Another advantage in using the upper floor is that the provision of specialized ventilation for chemistry in somewhat easier.

In areas where land costs are not excessive, there is a trend toward buildings of one story. Such building can provide the advantages of both the first and the top floors. Local considerations of building costs, the type of science work to be encouraged, and the facilities to serve this work should be the governing factors.

(3) *The Interrelation of Science Facilities.* Various principles concerning the interrelation of science facilities should guide those responsible for planning. Through careful application of such principles, the rooms and their equipment can be made more valuable to all who use them, and more economical investments are often possible.

In the elementary school there is a question as to whether science facilities should be located within each elementary classroom or should be placed in a separate elementary science room. There are certain psychological principles to be served in placing the major portion of the facilities for a grade in the the room that is usually used by that grade. For example, if the fifth grade facilities are located in the fifth grade room, the pupils have access to the science facilitics in the normal working situation. It is not necessary that the learners move to an unfamiliar situation with its distractions. Likewise, the teacher will probably make greater and more effective use of the resources because of the familiarity which results from everyday contact.

In some schools it has seemed more desirable to develop a separate science room to serve several of all elementary grades. In such instances overcrowding within the regular classroom may be alleviated, and the resources of the school may be more extensive because of the sharing of the facilities. In some instances it has been found desirable to combine the science room for the elementary school with a school museum, Many teachers regard this as a rather satisfactory arrangement. Such a museum is generally characterized as a *working museum.* It contains changing exhibits, contemporary as well as older materials. working models, study areas for pupils, and other encouragement and provisions for the participation of the learner.

In the high school the following principles should be kept in mind in determining the location of science rooms:

(a) The science rooms Major rooms and accessory rooms)

should be located near each other, with doors between the rooms.

(b) The rooms where the noise level may be relatively high should be separated from rooms where quiet is desirable.

(c) The rooms in which the activities produce dirt and debris should be well separated from the rooms where cleanliness is most desirable.

(d) In schools having several science rooms, as a large high schools, there are some advantages in a veritical grouping of science rooms. For example, the general science rooms may be directly above the biology rooms, and the physlcs rooms above the chemistry rooms. The advantages are in the reduced cost of plumbing, shorter flues for venting. and the possibilities of uniform deep design of rooms. The disadvantages are in tbe veritical travel, poor communication, and the general lack of interrelationships among the various subject fields of science.

(e) Shared facilities should be readily available to all who use these facilities.

(f) Each facility should be independently accessible. No facilty should be rendered inaccessible because of the use of an intervening area. This should include departmental or room liberaries.

(g) Certain facilities should be available from both the inside and the out-of-doors: for example, it is considered desirable for plant and animal rooms to be so accessible. School gardens should be easily reached. Materials and facilities needed for field trips, hikes on nature trials, and the like should be readily accessible.

(h) Science rooms should, in general, be relatively near those rooms used for crafts and home arts. The sharing of certain accessory rooms, such as those used for electronics and photography, make such an arrangement desirable. Such accessory rooms may be placed between the rooms used for science and those rooms used for other fields and should be accessible to both.

(i) It is desirable to place the scienee facilities relatively near the audio-visual center since the science teacher is likely to make extensive use of projected materials. The audio-visual center as a resource area for teachers (or small groups of students) is growing in favour. It houses materials and equipment, and provides for such preparation as previewing motion pictures and slides, making charts, and listening to recordings. Many of its resources can be used by the science teacher.

FACILITIES FOR STUDENT WORK

Furniture for Science Rooms

The selection of furniture for science rooms should be based upon the general nature of the work anticipated. Just as the plan for the room, and the use of out-of-door facilities reflect the quality of the science experiences, so should the furniture selections be made in the light of the needs of the total situation. Furniture should be evaluated only as a factor of a total situation. It is possible to cite the Probable effects of the use of given types of furniture or instructional procedures. The furniture necessary for laboratory work, reading, discussion, storage, demonstration, and display and the accessory fixtures and utilities should reflect an analysis of the functions they are to serve, and not a preconception gained from college installations or conventional school provisions. In general, the furniture should be adaptable to class, small group, and individual science work. The necessary furniture may come from commercial procedures, or, in some instances, may be locally provided.

Stock Units and Furniture Costs

Commercially produced furniture is based upon the use of

stock units of bases for laboratory desks, fume hoods and storage cabinets; rails and frames for desks and tables: legs for tables; and entire storage cabinets, Because such stock units are produced on a mass basis, their cost is relatively less than it would be if each item were individually planned and produced fo r each separate installation of furniture.

Each of the several companies produces a wide range of stock units from which the purchaser can choose in assembling furniture. The specification of furniture which cannot be assembled from stock units results in appreciably higher costs. Such specification should be avoided unless thc furniture needs are so clearly unique that custom-made items offer the only solution.

Wood and Metal Furniture Compared

Among producers and among users there are differing points of view with regard to the preferable material for furninure—wood or metal. The subject is controversial, with no clear cut decision. The following advantages are climed for wood furniture:

1. Natural warmth and beauty with chemically resistant clear or coloured finishes.
2. Quiet operation.
3. Ease of maintenance.
4. Natural resistance to chemicals.
5. Adaptability to alterations.
6. Natural insulating qualities protecting contents against minor fires.

The advantages claimed for metal furniture are:

1. Structural strength.
2. Fireproof—will not support combustion.
3. High ratio of usabie space to floor space occupied.

4. Permits high temperature baking enamels, developing maximum chemical resistance.
5. Non-absorbent to moisture, fumes of liquids.
6. Ease of operation of moving parts?

Variability of Furniture Assembly

The various types of demonstration and laboratory desks illustrated in Figures 6.1 to 6.6, have a given array of utilities. These services can be varied, using the basic furniture types. In a given situation it may be necessary to add compressed air or hot water to the utilities normally provided, or the electrical outlets may be omitted. Variations in the utilities generally involve added expense. The changes should be anticipated at the planning stage rather than attempted after the installation is completed.

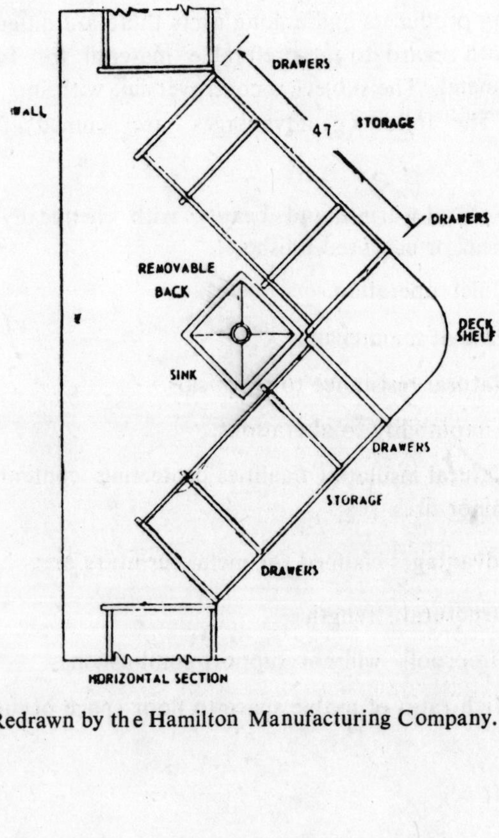

Redrawn by the Hamilton Manufacturing Company.

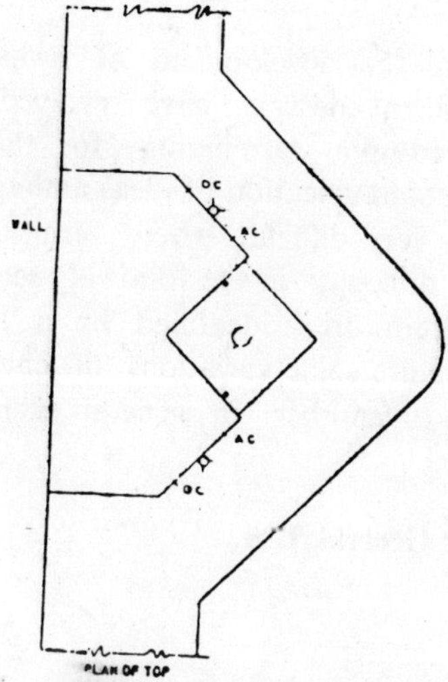

Pittsburgh Public Schools

FIG. 6.1: Multiples of this wall table can be planned to meet the needs of each science room. A view of this unit is shown in Fig. 6.1

In the planning of demonstration and laboratory desks, storage cabinets and other furniture, reference should be made to the various stock units that are available. The storage provisions in the desks can be varied considerably by the choice of cabinets that have various numbers and sizes of drawers, shelves, and other storage spaces.

The pattern of assembly of the various basic units in items of furniture can be altered as the local situation requires. For example, the assembly of eight-student desks from stock units of bases. The plan for this "island desk". Variations of wall tables assembled from stock units are shown in Figure 6.1. The table top is cut from stock to fit the resulting assembly of base units. Through such assemblies one school system is meeting the needs of the science laboratories.

Types of Laboratory Furniture

In the course of the development of furniture for science rooms many different designs have evolved. Unfortunately there is no agreed-upon terminology for the various types; consequently the communication of ideas among those responsible for planning is very difficult where furniture is concerned. To obviate this difficulty in the following sections each of the different types of furniture is identified by a letter. While for any one type there are some variations of characteristics from one manufacturer to another, in general such differences are minor.

Science Desks for General Use

Type A (Fig. 6.2)

This desk is half of the four-student desk in Type B, thus twice as much plumbing is required for the same number of students. Type A has the advantage of making a one-way facing room. It is obtainable (from different sources) in either 30 in. or 36 in. height and with or without attached chairs.

Type B (Fig. 6.2)

The orientation of this type of desk with respect to the room is variable. It can be so placed that most students with a slight turn face the front or it can be used with half of the students not facing the front of the room. If the desk is to be used only in a laboratory and not for class purposes such facts of orientation are not so significant,

Type C (Fig. 6.2)

This desk is essentially an extension of Type A; it is produced also for six students. It provides for a one-way facing room.

While all desks with utilities make rooms less flexible, the longer the combination of desks, the more rigid the room becomes, This desk is made with heights of both 30 and 36 in., and is obtainable with attached chairs.

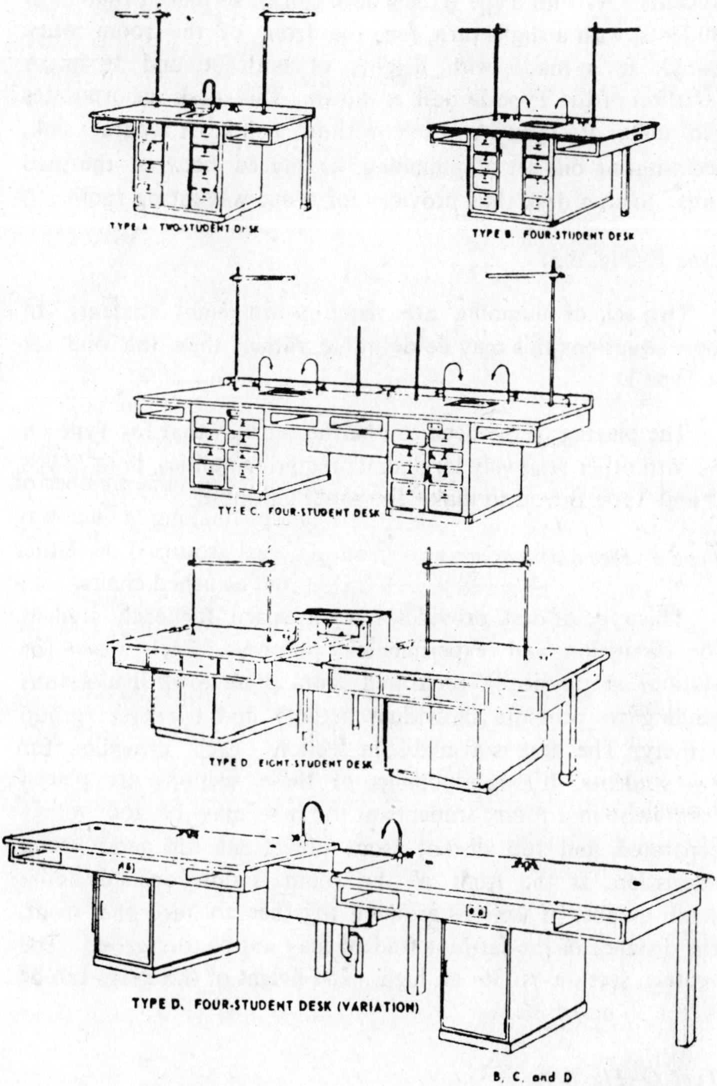

FIG. 6.2. Science Desks for General Use—Types A, B, C and D.

Type D (Fig. 6.2)

With this type of desk one set of plumbing serves eight students. As with Type B, this desk can be so placed that most students, with a slight turn, face the front of the room when seated. It is made with heights of both 30 and 36 in. A variation of the Type D desk is shown. This desk incorporates two units of Type A, each without a sink. A separate sink, necessitating one set of plumbing, is placed between the two units. Such a desk thus provides for a one-way facing room.

Type E (Fig. 6.3)

Two sets of plumbing are required for eight students. In some situations this may be desirable, rather than the one set in Type D.

The placing of the desk can be the same as that for Type D. As with other relatively large units requiring utilities, both Type D and Type E tend to make the room inflexible.

Type F (Fig. 6.3)

This type of desk provides a work space for each student for discussion and experimental purposes. Each space (or station) is physically separated from other work spaces, thus tending to promote individual activity and to deter group activity, The desk is available in sections, each providing for two students. If long groupings of these sections are placed lengthwise in a room, students at the rear may be too widely separated and too distant from the front for good group discussion. If the front of the room is along one side and a small number of sections grouped together to face the front, the distance of the farthest student may not be too great. The central section is 36 in. high. The height of the wings can be either 30 in. or 36 in.

Type G (Fig. 6.3)

The wall table provides for work space that must be fixed

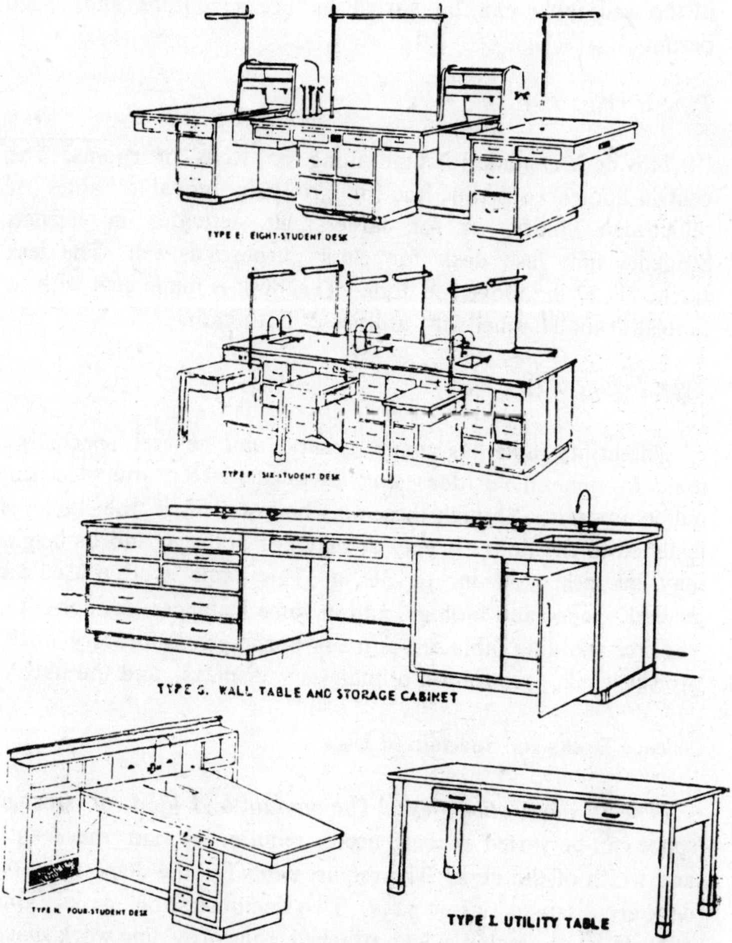

FIG. 6.3. Science Desks for General Uses—Types E, F, G, H and I

because of utilities. It can be profitably used in a science room in which it is not essential to have all utilities at all stations or in a room in which one or more free-standing desks provide for a portion of the stations. Various uses may be made of the space above the wall table: it may be left free, it may have tackboard for display, it may have storage cases, or a fume hood may be mounted on the wall table. The length of the wall table can be varied as needs require and space permits.

Type H (*Fig. 6.3*)

This desk is made to stand along the walls of rooms. The central floor area is thus free for grouping movable tables or tablet-arm chairs, or for large scale activities in science. Students may use desk for class purposes as well. The desk surface is 37 in. above the floor. This desk is made also with a central drain channel with utilities at both ends.

Type I (*Fig, 6.3*)

The utility table has many general and several specialized uses. In general, it does not have gas, water, and electrical outlets upon it, although these may be installed if the table is fastened to the floor. It may or may not drawers, and its height may be either 30 in. or 36 in. This table is often used for general science and biology, and in some instances, for physics. The free-standing table serves many purposes: laboratory work, project work, preparation of materials, displays, and the like.

Science Desks for Specialized Uses

The Demonstration Desk. The amount and kind of storage space can be varied as local needs require, as can the length and width of the desk. The requirements for the demonstration desk are discussed latter page. Two demonstration desks. One of them has a teacher's desk attached, thus providing work space for the teacher; however, this design limits thc flexibility of arrangement and action.

General Science and Biology Tables. The separate table for for general science and biology is often a general purpose table or utility table. The type shown in Figure 6.5 is quite satisfactory where little storage is necessary. Such tables without utilities provide flexibility in the use of the seating area of the room; this presupposes that the necessary utilities are avail-

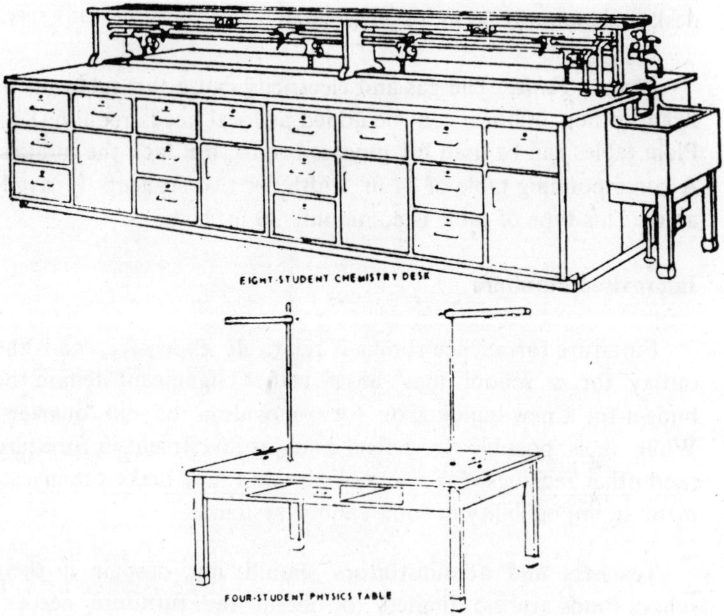

FIG. 6.4. Science Desks for Specialized Uses: Chemistry, and Physics.

able elsewhere. These tables are commonly 30 in. high. The length varies from 54 in. to 72 in.

Biology Desk. This tape of desk provides storage space within the desk. The space for storage is available in various arrangements of drawers and cupboards. This type of desk for four students is commonly 42 in wide: it is made in a 24 in. width for two students. The latter variation provides for the ready use of this dek in a one-way facing room, although the four-student desk can be used (refer to Type B). This type of desk is generally 30 in. high.

Chemistry Desk. The type of desk shown in Figure 6.6 is, most appropriate in the room set aside for laboratory use only, although one or more such units can be placed in a classroom laboratory or a multipurpose science room to provide extra stations where most stations are provided by wall tables. This desk is commonly 36 in. or 37 in. high.

Physics Table. The gas and electricity cause this table to be fixed; other utilities can be added as local needs require (6.4). Plain tables can be used for physics rooms, but lack the utilities. A corresponding table of 24 in. width for two students is available. This type of table is commonly 30 in. high.

Improvised Furniture

Furniture for science rooms is relatively expensive, and the outlay for a school may be a rather significant item in the budget for a new building or for renovation of old quarters. While it is possible to defend a major investment in furniture (and other facilities) for a school, the cost may make the investment an impossibility in some school systems.

Teachers and administrators should not despair if their school funds are too limited to obtain the furniture needed. While there are limitations in the use of improvised furniture, it is possible to provide many worthwhile experiences with tables, benches, shelves and cupboards originally designed for other purposes, or built from inexpensive materials at hand.

In the absence of appropriate furniture, the following measures can be taken to provide at least a minimum of laboratory, demonstration, display and storage furniture;

1. Obtain unused furniture from other parts of the school.
2. Use such available tables and cabinets as can be obtained from community sources. These can be adapted by acid-proofing the working surfaces, repainting or revarnishing other surfaces, changing the shelves and drawers and altering the height.
3. Build a demonstration platform to be placed upon the teacher's desk. Such an addition provides the desired height at little expense.

4. Build work tables or benches from boards and planks. Industrial arts classes may cooperate in such a project and, with care and effort, fairly satisfactory working surfaces can be provided. The top can be acid-proofed.

5. Shelves for storage and display can be built from locally obtained lumber.

Utilities for Improvised Furniture

Attention should be given to the services for improvised laboratory desks. It is not always practicable to provide running water on an improvised desk, but it may be possible to place such a desk adjacent to a water faucet and sink. It may be necessary in some cases to provide the running water by siphon from a tank supported above the desk.

Electicity can be provided by using conduit or flexible cable with outlets at appropriate points. Care should be exercised to follow code requirements in making such installations.

Gas lines should be extended to gas cocks above the work tables. The lines should be carefully tested before use. If natural gas is not available in the school, gaseous fuel liquefied under pressure can be readily provided in most communities.

Placement of Fixed Furniture

The use of utilities on desks in science rooms requires that the desks be more or less permanently placed. The placement of such desks along walls helps to avoid the inflexibility that results from fixed furniture. Where it is essential that the furniture be fixed, the following recommendations for its placement have been found to be satisfactory:

1. The demonstration desk should be centered at the front of the room at a distance of approximately 3 ft. from the front wall.

2. The front row of student laboratory desks should or-

dinarily be at a distance of 3½ or 4 ft. from the demonstration table, although there may be occasion to place them closer than this distance.

3. A space of approximately 3 ft. should be allowed between each two rows of desks from front to back of the room and approximately the same distance between the rows of desks across the room.
4. The aisle on the window side of the room should be wide enough that aquaria and germinating flats can be placed in front of the windows with at least a three-foot width remaining.
5. Short desks are more practical than long ones because they allow easy passage from row to row.

Facilities for Demonstrations

Successful demonstrations require a satisfactory demonstration desk with the necessary facilities. There must be also a minimum of demonstration equipment, whicb, is, in general, relatively large and simple.

The demonstration desk should be high enough for students to see the demonstrated materials easily from their seats. Heights are commonly 36 in. and 37 in, The desk should be well-lighted: supplementary illumination may be necessary. Electrical outlets provide for a variety of uses, including movable shaded lamps for the desktop, The demonstration desk should have as a part of its cabinet work adequate storage for commonly used materials. Running water and gas should be provided on the demonstration desk. A useful accessory to the demonstration desk is the cart or table on wheels. Sueh a table, supplied with rubber-tired casters. can be used in a variety of ways: for prior preparation of demonstrations (in the storage room, for example). as an extension of the demonstration table when it is not large enough, for auxiliary demonstrations and displays in various parts of the room, and in classrooms where there is no demonstration desk. The dimensions are optional, although a height equivalent to that of the demonstration desk with which it is to be used is preferable.

Facilities for the Science Library

A library and reading facilities should be an integral part of of the classroom-laboratory. Open shelves or an enclosed cabinet provides for books: open shelves are more inviting, but an enclosed cabinate furnishes more protection from dust and dirt. A table and comfortable chairs should be provided. In the usual school such classroom reading facilities are needed for only a few students at any one time, but are none-the-less essential.

The resources of this library include books, magazines and pamphlets. The books are classified and catalogued by the central lihrary of the school. The circulation of the books is handled under the direction of the science teacher. The books are of two general kinds; first there is a so-called permanent collection of reference books, textbooks, and pamphlets in the fields of science that the classroom-laboratory serves; a second group includes volumes temporarily removed from the central library of the school to be used for short periods in connection with a particular study.

The magazines provided are of both general and specialized scientific nature, although the usual library has little use for the more technical magazines. A shelf or rack for their display is essential. Bound volumes of past issues of magazines are housed in the central library.

A pamphlet file is needed so that this type of resource can be readily used and kept alive. In many cases this kind of material is the best source of recent information. The filing and storage of pamphlets requires continuous attention. Various methods are used. Some teachers place each pamphlet in a filing folder, which is titled and placed in the drawer of a filing cabinet. Others use pamphlet binders, placing them on shelves. A third system utilizes filing boxes, each box accommodating many pamphlets. Storage provision is essential for duplicate copies of pamphlets. In some instances the teacher has sufficient copies that one is available to each student. Such duplicates may be tied together, or placed in large envelops, and housed in enclosed cabinets or drawers.

Storage of Apparatus and Supplies

Separate storage rooms for scientific apparatus and supplies are desirable in most schools. Often such rooms are used also for preparing for demonstrations and laboratory work and in the repair and adjustment of apparatus. Where complete darkening is possible the storage room may be used also for photographic work as well as for experimental to various kinds in connection with science courses.

In planning for a separate storage room, three factors should be considered:

1. It should be located next to the room in which the largest number of stored articles will be used.
2. Double doors to the storage room make it more accessible. A door from the corridor into the storage room is also an advantage, especially if it is located in a science unit of several rooms,
3. Forced ventilation should be provided.

The separate storeroom has advantages and disadvantages, The advantages are:

1. It leaves more space for work centers in the classroom,
2. One lock and key controls the storeroom.
3. It provides space for equipment repair, projects, dark-room work and other activities,

The disadvantages are:

1. Equipment stored in one room may not be convenient to all work centers.

2. Storing the supplies and equipment of individual students in the storage room may result in crowding when they are taking out and returning their materials.

3. A separate storage room is sometimes difficult to keep in order since materials may be dumped there awaiting the final storing.

These advantages may be minimized by careful planning and supervision.

Storage in the Classroom-Laboratory

There should be provision for the storage of equipment and supplies in the classroom-laboratory. The amount and kind of provision depend upon such factors as the amount and kind of storage in a separate room, the number of students to be served, and the nature of the apparatus and supplies to be stored. In the interest of time saving and the full use of resources, the storage of such materials should be, whenever practicable, in the room in which they will be used. Separate storage cabinets placed in different parts of the room prevent congestion in obtaining and returning apparatus.

The anticipation of storage needs in the planning of science rooms is of considerable importance. The recessing of storage cabinets into the walls of the room provides several advantages over the use of separately built, free standing cabinets:

1. The floor space required for free-standing cabinets is saved,

2. The cost of the cabinets is included in the initial cost of the building and should be less than that of construction under special contract.

3. The appearace of the room is improved.

To profit by these advantages it is essential that the storage space be planned as the building itself is being planned. Because recessed cabinets cannot be shifted after the building has been

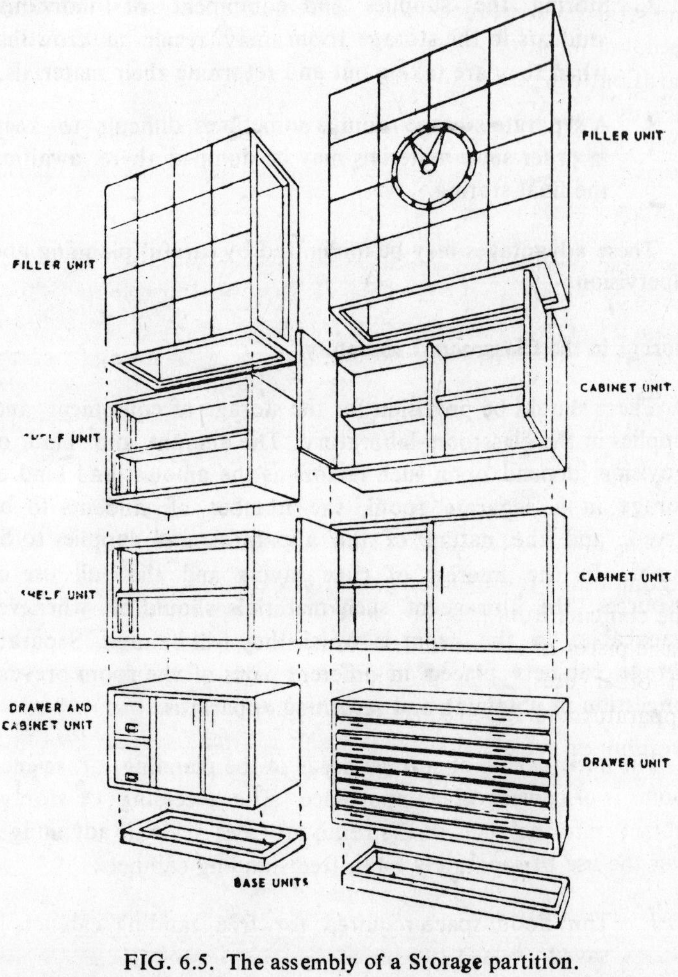

FIG. 6.5. The assembly of a Storage partition.

completed, their location and design must be determined with care. In this respect the free-standing cabinet is somewhat advantageous: a limited change of its location is possible.

A type of storage facility that has some of the advantages of both the built-in cabinet and the movable cabinet is the storage wall. As indicated in Figures 6.15, the storage wall is assembled from units that fit together. The units are of various designs and serve different functions; thus they are chosen to meet the needs of a given situation. Such a facility can be used as a partition to separate two rooms or can be assembled along an already existing wall.[3] Accessory rooms for such purposes as darkroom work or preparations can be provided by separating a portion of the classroom-laboratory by means of such a storage wall.

It has been found desirable and practicable to provide two types of storage facilities. One of these is the general storage room or storage cabinet in the classroom-laboratory accessible to and used by the students. A second type of storage facilities is a cabinet or small closet that provides for those materials that must be kept under very careful supervision. In some situations locks are essential.

Where a room is to be used by different groups, separate storage facilities for each group are necessary. This may be accomplished through the use of tote drawers, used to transport the contents from place to place. Such drawers may contain basic pieces of apparatus, all drawers being similarly stocked. In other uses, each drawer may contain an assortment of apparatus and supplies needed for a given experiment, demonstration, or project: a series of such drawers thus can store the materials needed for a variety of activities. Cabinets with such drawers may be used as partitions to separate a room into specialized work areas.

Storage of Charts and Maps

Charts and maps require specialized storage provisions. Without adequate facilities these resources soon become unattractive and unusable. Because of the large number of such materials and the variety of sizes, there seems to be no simple solution. Various methods are used, each with its advantages. Among the provisions that have been used satisfactorily are the following:

(a) Blueprint cabinets having long shallow drawers with several charts in each.

(b) Long pigeon-hole shelves with rolled charts in each.

(c) Sets of charts (generally commercially produced) bound together at the top by a charthead and supported on a tripod.

(d) Suspension of charts from a rack which is drawn up by means of pulleys to a position near the ceiling.

(e) Suspension of rolled charts from racks which can be extended.

(f) Suspension of flat charts from hinged panels, providing for both storage and examination. These hinged panels can be approximately 4 ft., long and thus accommodate relatively large charts. Spring clothespins hold the charts in place. A chart cabinet of smaller size and providing for fewer charts is described as follows:

> "This cabinet contains five hangers, speed approximately 2½ in, apart, hinged at left to swing forward. The hinged hanger is a means of making a chart more accessible. Each hanger has two squarebend hooks placed 12 in. apart on which to hang charts. The cabinet can be made of 1/4 in. plywood. The sides and doors are reinforced with a framework of material 1 in. by 2 in. The top and bottom should be of 3/4 in. plywood.
>
> The hangers are made of 3/4 in. plywood."

Charts on relatively thin paper and without metal ferrules should be reinforced across the top in preparation for this type of storage.

(g) Suspension from bars in a large compartment. Such a bin is provided as a part of the base of a storage and display unit. A separate section provides storage for industrial exhibits, filmstrips, pamphlets, and magazines. Shelves above provide for apparatus storage. Sliding panels of chalkboard and tackboard provide work and display space.[7]

The method used should depend upon the needs and resources of the individual school. Whatever method is used, it should make charts and maps readily available and give them protection in storage.

Temporary Storage

Special consideration should be given to the provisions which should be made for the temporary storage of textbooks and other materials carried by high school students from one class to another. If planned storage for these materials is not provided, they are often stacked on the work-surfaces, or on the floor, where they are subject to damage, impede movement, and monopolize space which should be used for other purposes. Some suggested locations for such storage are:

1. Open shelves near the entrance to the room, or on the end of the demonstration desk.
2. Book compartments, or racks, in the student tables.
3. Racks built into student chairs.

GENERAL PROVISIONS FOR SAFETY

Experiences in science involve the use of many pieces of apparatus working with materials, and using energy. Often the experience involves motion, either of the student of the equipment. In all these instances and many others there are potential hazards to safety. Provisions that promote safety in science rooms should be anticipated as the science facilities are being planned. Those provisions of general nature include the following:

1. Flooring that is not slippery when wet.
2. Furniture with rounded corners and edges.
3. Wide aisles and exit passages.
4. Cabinets fitted with shatterproof glass.
5. Prominently placed shut-off values for gas and water, lines.

6. Master switches for electrical circuits.
7. Placing of electrical outlets on pedestals or splashboards to prevent entrance of water by flooding.
8. Means of prompt removal of any explosive, toxic, or obnoxious gas.
9. Study chairs, tables, and other furniture.
10. Approved fire extinguishers and fire blankets. Care must be exercised in the choice and use of fire extinguishers. The carbon dioxide type is of most general value. The carbon tetrachloride extinguisher is good for fires in electrical equipment but is dangerous when used where the ventilation in inadequate.
11. Carefully planned storage space, with respect to such qualities as:
    (a) Ready accessibility, so that apparatus can be easily and quickly returned to place.
    (b) Adequate amount so that damage and danger do not result from crowding.
    (c) Considerable depth of shelves so that articles do not become readily dislodged.
    (d) Avoidance of high storage space, but provision of a strong stepladder or stool with wide treads if there must be high storage.
    (e) Use of racks, or drawers with partitions, for sharp tools.
    (f) Special storage for chemicals that are highly toxic inflammable, or that must be kept covered with liquid.
12. One or more first-aid kits in any science room.

AUDIO-VISUAL FACILITIES FOR SCIENCE TEACHING

The teaching of science is so intimately related to and so

dependent upon the material aspect of the universe that the inclusion of a wealth of the real things—samples, specimens, pictures, models, and other representations—is a necessity for effective results. The use of visual facilities to present otherwise abstract relationships and unobservable motion is an advantage to teaching accepted almost unanimously by teachers. The motion pieture, radio and television bring inaccessible scenes and events to the learner. Not the least of the values in the use of these facilities is the participation of the learner in their preparation and use. The preparation of displays, giving of demonstrations, and the making of charts to be used in reporting are effective methods of providing for student use of such resources. The good science room provides an extensive range of audio-visual facilities.

It is essential that the presence of audio-visual resources not not be confused with their effective use; nor should more mechanical manipulation be so regarded. However, use is not possible unless provision is made for facilities of the various kinds available. Certain of the resources must be purchased, and others are free; some are the real materials and specimens, some are representations: some are in themselves applications of principles to be comprehended, others are accessory devices which make possible the use of learning materials. Regardless of the form or specific function of such facilities. their effective use is predicated upon their accessibility to the teacher and the presence of such essential accessories as display areas, tack-boards, projection screens, and well-darkened rooms. Such conditions are of general nature, being equally pertinent to the rooms used in an elementary school, the various classroom-laboratories provided for high school science courses, and corresponding rooms used in the education of science teachers in institutions for teaeher education.

The general provisions for audio-visual materials in a room for general science do not differ greatly from those in a room for physics, biology, or chemistry; the specific provisions do vary in many respects. Such variations are discussed in the appropriate chapters.

Types of Audio-Visual Facilities

Visual materials and facilities can be classified broadly into two categories: non-projected. Resources utilizing sound may be considered in a third category. Each type has its particular values and advantages. The science room or rooms should have, except in rare instances, a generous array of resources from each of the three categories; without such a distribution of facilities, the opportunities for effective work are limited.

1. Non-Projected Facilities:

A. *Objects, Specimens, Models*

The real things of science are of great importance to teachers, and adequate provision should be made for their display, use, nnd storage. Specimens, objects, samples, models, and mock-ups are among the most common of resources—they are near the learner's experience. In many instances they are used in an experimental situation in a laboratory; in others they are more effectively used for display. In a very real sence the apparatus, equipment, and supplies used in an experimental situation fall within the classification of multi-sensory aides. Such facilities and their uses are separately considered in the various chapters.

Adequate space for display and exhibit merits careful planning. Two general devices are used: open shelves, and exhibit cases. Wide shelves strategically placed in the room may be used to display many appropriate and interesting materials, Including the on-going work or completed projects of the students. Such a shelf should be located higher above the floor than the standard desk height of 30 in. The height of the students concerned is a factor; shelves at distances of 36 in., to 72 in., above the floor have been satisfactorily used.

The exhibit case is an asset of considerable potentiality to the science teacher. It is quite desirable to have a science exhibit space or case available to all students in the school. Such a case may be placed in an alcove of corridor; in some schools such a case has been built into the corridor wall, with exposure to the carridor and access by means of doors in the rear of the case from a science room. A separate exhibit case in the science room is desirable. Such cases can be built in the school or purchased. Some types of display case from stores have been used satisfactorily.

The arrangement of the exhibit material and the illustration of the space should be done with care. In general, the exhibit case should have adjustable shelves and may very well provide for more than one level of display. The back of the cabinet may be of backboard so that explantory material can be readily attached. Exhibits in science are such that a relatively large cabinet should be provided. Minimum dimensions should be a width of 4 ft., a height of 3 ft., and a depth of 2 ft. The illumination of the case map be accomplished by either incandescent or fluorescent units. The case should contain an additional electrical outlet so that models requiring electric current can be operated. Ventilation should be provided to permit the display of live animals.

The principal object of display should be approximately at eye level. To the extent that real objects. specimens, and models can be so displayed. the exhibit proves interesting and informative. The back of the cabinet, preferably of a light colour, is used for explanatory material, diagrams, and drawings and pictures that lend atmosphere. The lower level of exhibit may con ain supporting and related material of some interest to most observers, but of a character secondary to the main exhibit.

There are many times when the object to be displayed is too large for either a shelf or a display case. In such instances the exhibit may be placed on a table or on the floor. The flexibility of room arrangement necessary for such exhibits should be anticipated in the planning of science rooms. The provision of open floor space in the room or space from which the furniture can be readily moved makes possible the use of large exhibits. If the total classroom-laboratory space is pre-empted by fixed furniture the lack of flexibility may prevent the use of such resources.

1. Non-Projected Facilities:

B . *Wall Display and Work Space*

The use of wall areas for display, directed study and illustration is of prime importance in science teaching. Chalkboard as a work and illustrative area is perhaps the most common facility. The traditional black colour is being gradually replaced by lighter colours which are more attractive and have higher reflection factors. Colours commonly used are light green, light yellow, and ivory. A convenient chalkboard for the front of the room consists of a six-foot panel of ivory glass board, flanked on each side by a panel of light green board. The light center section may be used for a projection screen, for drawing graphs and diagrams (in black chalk), and as a light background for demonstrations. While this arrangement has both utilitarian and esthetic values, it requires two colours of chalk.

Tackboards in the Science Room

Tackboards, or bulletin boards, are an essential part of the wall surface. Cork is a good material for tackboards, but certain kinds of vegetable-fiber board are also satisfactory, particularly for a semi-permanent or temporary installation. The same colors as those for chalkboards are desirable; most material can be painted the desired color. One of two panels in the room may well consist of reversible panels, chalkboard on one side and tackboard on other. The dimensions of the tackboard panel should approximate those of the chalkboards. As with panels of chalkboard. that are quite small tend not to justify their cost. The placing of a narrow strip of tackboard above the chalkboad is a questionable investment, Such a tackboard is satisfactory for the suspension of charts and similar materials that are low enough to be read or examined carefully; how, for the display of materials to be observed closely at a level considerably above the eyes it is rather unsatisfactory. The use of such a strip is a limited to narrow murals or other decorative material.

The position and construction of chalkboards and tack-boards in the science room should be planned for their most effective use. In general, they should be placed between windows nor on walls opposite those having windows: neither position is favourable for visibility.

Achieving Flexibility in Display

The need for varying amounts of chalkboard and tack-board, along with the availability of a limited amount of wall space, has led to the use of adjustable panels of various kinds, Two familiar types are those that are reversible and those that are held in tracks. One such familar device provides for the vertical adjustment of chalkboards, two or three panels being held within one frame. A projection screen can be placed permanently on the innermost panel, or a rolled screen can be supported on the upper part of the fame of the outermost panel, its vertical position thus being adjustable

Flexibility is promoted further through the use of portable panels of tackboard and chalkboard. The adaptation of a conventional classroom for use in science teaching is promoted through such resources. Special reports, projects, small group work, and science fairs generally require special display facilities: the portable panel is well adapted to such uses.

The familiar floor-stand tackboard or chalkboard is a useful display and work space. It may be fitted with a tack-board on one surface and a chalkboard on the other. In some schools, panels of tackboard of dimensions approximately 3½ ft. are equipped with screw eyes on the top frame and suspended from hooks at desired locations in the science room.

The horizontal adjustment of panels of chalkboard and tackboard can be provided by tracks placed on shelves along the front or side of the science room. A storage area for unused panels is provided at one end. Any panel is moved into positian as it is needed.

A display rail or map rail is a valuable addition to wall display facilities. Such a rail serves certain of the functions of a tackboard; a cork portion accommodates thumb tacks. Ajustable hooks support larger and heavier kinds of material. The display rail may be placed on the frame above the chalkboard, or at other strategic points in the room.

Display panels in the form of wings which can be adjusted or removed, are commercially available or can be made locally. Such panels of tackboard are commonly of dimensions approximately 2½ ft. by 3 ft. They are hinged or pivoted along one side; front and back can both be used. Several scuh panels in a series provide a generous display space with a limited use of wall area, Additional display space can be provided by replacing the door panels of storage cabinets with tackboards in those situations where the cabinets are in the classroom-laboratory. (Similarly, panels of chalkboard can be placed, in the doors of storage cabinets.) Whatever tackboard space is provided, attention should be given to its illumination. Supplementary lighting should be provided for any such work area that is not well-lighted.

(2) *Facilities for Projection*

The use of projected materials involves special attention in planning since projection requires special provisions. The problem of projection includes the important consideration of darkening facilities, electrical outlets, sound cable, screens, acoustic, and ventilation.

(a) Darkening Provisions. Several ways can be used to darken rooms; some are better than others. The practical solution is to choose the method that best fits needs and budget. Four economical methods of darkening rooms are: drapes, roller shades, Venetian blinds, and blackout shields. Drapes give a semi-dark or even a very dark room; black roller shades if boxed at the sides, create nearly total darkness; shades that much the interior colour scheme produce semi-darkness as do Venetian blinds. Blackout shields of 1 in. Celotex or 3/8. plywood provide nearly total darkness. (If a room is completely darkened, provision should be made for a dim light to be placed behind the screen during projection to reduce the eyestrain reulting from the brightness ratio between the screen area and its surroundings.)

A somewhat more elaborate provision for room darkening has given satisfactory results. Light-proof drapes are hung from a track at the ceiling and extend the entire length of the wall. The track is placed approximately 18 in. from the wall, and the drapes are of such length that the lower edge is approximately 12 in. from the floor. The room is easily and effectively darkened. In those climates were supplementary ventilation is provided by open windows, the air can move beneath the draps. The degree of darkness required depends upon the nature of the projection. Microprojection, opaque projection, and coloured slides and filmstrips require well-darkened rooms. Partial darkness is generally adequate for lantern slides, black and white filmstrips and motion pictures. Assuming a degree of partial darkness, the teacher can use projectors with bubls of relatively high power and obtain satisfactory results. The provision of a shield or hood to protect the screen from interfering light makes possible so-called daylight projection.

Provisions for darkening the classroom-labortory should be related to plans for a photographic dark-room. In some schools there is no separate provision for a darkroom; instead, the classroom-laboratory is fitted for complete or near-total darkness. Such arrangements provide also for projections, experimental work in optics, studies of fluorescence, and the like. Schools of smaller and moderate size may find such planning advantrgeous, although it has the disadvantage of limiting the use of darkroom facilities when science classes are in session.

(b) Ventilation of Darkened Rooms. Adequate ventilation of a darkened room is sometimes difficult to accomplish. If forced ventilation is inadequate, exhaust fans can be mounted in windows, or lightt-rap ventilators may be installed. In some situations air ducts can be built into the wall beneath the window. If the classroom is to be darkened frequently, a window-sill air conditioner should be considered. Open windows for supplementary ventilation can be used with drapes, as previsously described.

(c) Screens. Many types of projection scree can be purchased at nominal cost. The width of the screen should be at least one-sixth of the length of the room. A long, narrow room can be fitted with a beaded screen to advantage, but this type is not suitable for a wide room. A beaded screen requires a viewing angle of less than thirty degrees for most satisfactory results. Ordinary white window shades are suitable and are often used; sign cloth is satisfactory and inexpensive substitute. A panel of ivory-tone glass chalkboard can serve as a fixed screen at the front of the room. If a fixed screen is uscd, a portable screen should be available for those instances when the seating arrangement is changed.

(d) Acoustics. If the accoustics are poor in the science room. the application of acoustic tile or board to selected portions of the walls and ceiling minimizes this condition. As a temporary measure, monk's cloth or burlap hung over cabinet, chalkboards, and walls is helpful. Mounting the load-speaker on a shelf above the screen may improve the acoustics of the room.

(e) Electrical Circuits. The science rooms should be equipped with electrical outlets in various locations to serve projectors (as well as other electric service needs). The availability of several outlets avoids the use of long extension cords. Planning for the science room should include a built-in speaker cable from the best location for the loudspeaker to a point near the best location for the motion picture projector. This cable avoids the necessity of placing the speaker cable on the floor where it may be damaged or may lead to accident. The light controls for the room should be easily accessible to the projectionist. Because the motion picture projector is generally used near the the rear of the room, a set of controls in this area is a desirable convenience. Overhead projectors for lantern slides are often used on the demonstration desk; a second set of controls for the overhead lights may be provided at this point.

(f) Projector Stands. The scienee room should be equipped with sturdy tables or stands for the support of projection equipment. Such tables should be high enough that the light from the projector to the screen is above the heads of the students. A desirable provision for the heavier projectors is a table on wheels. Such a table serves to transport the projectors from one room to another and makes adjustment of the position of the projector easy. A drop-leaf table attached to the wall is a satisfactory support for the projector. The equipment is somewhat protected, but there is a loss of flexibility.

(3) *Auditory Facilities*

In rooms used for radio listening, projection of motion pictures, the making and using of recordings, and the viewing of television programmes particular attention should be given to the acoustics, as well to the needed equipment and materials.

The availability of radio programmes of scientific nature

makes a radio receiver a valuable asset. In many cases, however, the radio programme is presented at a time not convenient to the school schedule of class; it may be given in the late afternoon or evening, during the weekend, or during a vacation period. While the school schedule should provide for needed variations to include such resources as good radio programmes, such changes are not always possible. A recorder (tape, wire to disk) bridges the time gap between the radio presentation and its use in class. Combination radio-recorders are available; in general, the same devices play the programmes back and may be used as public address systems as well. Such equipment should be readily available to the science teacher.

Facilities for television prorammes should be anticipated in planning the science facilities, even though immediate installation is not the objctive. A conduit or other channel should be provided at the time of bluilding to accommodate the television cable to the science rooms.

In the usual situation it is desirable to plan to share projection, recording, and playback equipment among the various teachers having use for it. In some schools all such equipment is administered through the adudio-visual center. This center has often taken the form of a separate audio-visual aids room, to which classes go for use of the equipment. Aside from conflicts in scheduling, the use of such a room seems undesirable from a psychological standpoint. Present thought favours the use of audio-visual materials in the normal learning situation, with the audio-visual center for preview and preparatory work.

Providing a Darkroom

A darkroom to serve the science laboratories as well as the entire school should be provided with facilities for developing and printing photographs. This room should be accessible from the classroom-laboratory, the preparation room, and the corridor. The space utilized for the darkroom varies from one school to another. If the classroom-laboratory can be well darkened, many simple but fundamcntal operations in photography and optics can be carried out and a separate darkroom is not necessary. With appropriate planning the darkroom can be combined with the preparation on the storage room. The minimum area for the darkroom should be 100 sq. ft.

Where a separate darkroom is provided, attention should be given and the access door. It should be possible for students to enter to leave the darkroom without disturbing others who are working there. Satisfactory methods of accomplishing.

One lighting fixture provides sufficient light for the general illumination of the usual darkroom. The switeh for this fixture should be far enough removed from the access door that the light cannot be turned on inadvertently by one entering. The use of such filters as the ruby light is determined by the size of the room and the operations to be accomodated. In the usual darkroom one such source over the work area is adequate. One or two duplex outlets for 110-v.a.c. should be provided for the use of enlargers, print boxes, and the like.

It is desirable but not essential that the walls of the darkroom be painted with a dull (non-reflecting) black paint. The paint is most needed on the areas of the access passage and doors, thus reducing reflection into the room.

The equipmcnt of the darkroom is directly related to the type of activity that it is to accommodate. What this varies from school to school, a few basic provisions should be made:

(a) The developing table should be provided with a sink, hot and cold water, drain, gas and a 110-v. a. c. outlet. It should also provide for drying films and prints. A cupboard and drawers provide storage space.

(b) There should be a wall cabinet above the developing table to provide storage space for and ready access to the necessary photograph supplies.

(c) A printing table with drawers and cupboard should be placed adjacent to the developing table. Light-tight drawers for printing papers are desirable. Additional work space for other equipment (such as an enlarger) can be provided by a utility table or a wall shelf.

(d) The darkroom is often used for experimental work with light and optics. To serve such uses, adequate work table space as well as separate storage compart ments for the materials used should be provided.

Public Programmes and Exhibits

The place of science in school assembly progrommes and in public exhibits should be taken into account as school buildings are planned. Such programmes and exhibits are commonly presented in the school auditorium or gymnasium; in many schools the auditorium and gymnasium are combined. The programmes and exhibits occur in various forms, such as:

— Demonstrations by science classes or clubs for school assembly programmes

— Programmes by science classes or clubs for parent night or open house

— Science days or science fairs for county, city, or district in connection with such activities as the Junior Academy of Science

— Travelling industrial programmes and exhibits

The major provisions for such activities to be anticipated in the planning of buildings are the utilities. Tables and other display facilities should be provided in relation to the needs of each school.

The stage of the auditorium should by supplied with lines for running water and drain, gas, and 110-v., a.c. The wiring for the electricity should provide for 30 amps. The gymnasium (or other large exhibit room) should be equipped with the same utilities along at least two walls. The number of outlets depends upon the anticipated needs. Experience indicates the desirability of several well-distributed sets of utility outlets as a minimum.

In the interest of safety all outlets should be enclosed in locked compartments to prevent tampering.

FACILITIES FOR THE SCIENCE TEACHER

Some otherwise superior school buildings are deficient in the facilities provided for staff members. Faculty rooms may be adequate, but often little thought has been given to the activities of the teacher. A science teacher regularly engages in a number of activities for which facilities should be provided; keeping records, making plans, preparing demonstrations, repairing and creating equipment for student experimentation, conferring with students, supervising student activities, assuming the responsibilities of a home-room, and attending to his own personal and professional growth through reading and other activities. The minimum equipment which should be furnished to each teacher is a desk and two chairs, a vertical file with lock, a book shelf, a card file, and a conference area. The preparation center, a major need of each teacher.

The teacher's desk should be separated somewhat from the classroom. Complete separation map be undesirable in view of the supervision needed in the usual situation. A possible solution is the use of a glass partition to surround the teacher's desk and file.

Where separate office space is not provided, the library corner or the preparation room may be used for conferences. This practice limits the availability of these facilities for their proper activities.

CHECK LIST OF GENERAL SCHOOL FACILITIES FOR SCIENCE INSTRUCTION

General Planning of Science Facilities

1. Is the building site selected, in part, because of its potential contributions to science teaching?

2. Do the provisions for the use of community resources indicate that science is an activity of both community and school?

3. Are the ideas of many qualified persons including those of the science teachers utilized in the planning of the science facilities?

Characteristics of Rooms for Science Teaching

4. Do the plans for science provide at least 35 square feet of area for each student in the largest group to be served in the science room?

5. Are the unique requirements of science upon floors. illumination, heating, ventilation, plumbing, gas and electric services, anticipated in the preparation of plants?

6. Do the plans provide that science rooms and furnishings shall have pleasing designs and attractive colours?

7. Can the rooms planned for science be increased in size or partitioned, and can adjoining rooms be readily converted to science rooms with all necessary facilities?

Facilities for Student Work in Science

8. Can the desks, chairs, tebles and the like be re-arranged for a variety of group, individual, and class activities?

9. Do the rooms provide for effcetive maintenance and use of an abundance of living and non-living materials?

10 Does the science room provide a special place where experiments may be demonstrated for and readily seen by the students?

11. Do the science facilities provide sufficient space for students working alone or in pairs to do experimental work?

12. Do the science facilities provide a place with adequate tools and supplies for the construction and repair of science apparatus and equipment?

13. Are there places in the science rooms where students and the teachers can carry on experimental work, arrange and try out demonstrations, or engage in junior research, without danger of the experiments being disturbed during the time when the work is in progress?

14. Is there library and reading space in the science rooms and is there an adequate section in the general library where stud^nts and teachers can read as they study and plan?

15. Are there rooms, shelves. cabinets, cases, and other arrangements for adequate storage for all items that have a place in effective science teaching, including those that are fragile, corrodible, or dangerous?

16. Is the storage space to distributed that materials can be readily obtained and returned?

17. Are the provisions for safety adequate with respect to injury by such potential hazards as furniture. utilities, chemicals, and fire?

18. Are first-aid supplies quickly available at any needed place?

Audio-Visual Facilities for Science Teaching

19. Are the rooms planned for science well adapted to the use of a variety of audio-visual aids, and is there an adequate supply of such aids accessible to the science teachers?

20. Do the science rooms, school corridors and other strategic places provide for the effective display of student and teacher prepared exhibits and products of science available commercially?

21. Does the science room, together with other school facilities, provide the materials and settings needed to produce posters, charts, radio presentations, news accounts, and other mass media that can be used to carry science to the whole school and community?

Facilities for the Science Teacher

22. Do the science arrangemenss provide a place for each science teacher to work on teaching plans, supply orders, achievement tests, records and reports?

23. Do the science rooms provide a place where the science teacher can confer with or work privately with students, parents, and others?

BIBLIOGRAPHY

1. Better School Buildings for Alabama, a Planning Manual, Bulletin 1950, No. 3. Department of Education, State of Alabama, Montgomery. (J.W. Letson and others.) October, 1949.

2. Coleman, H.S., (Ed.), Laboratory Design, New York: Reinhold Publishing Corp., 1951.

3. East, Marjorie and Edgar Dale (Ed.), Display for Learning, New York: The Dryden Press, 1952.

4. Elliott, William P., "Are you Planning to Renovate Your Science Laboratories?" School Science and Ma hematics 42: 35-38, January, 1942.

5. Hart, F.W. and Willard E. Givens, "Science Laboratories in Secondary Schools," American School and University, Third Edition, pp. 435-438, 1930-31.

6. Herrick, John H., et al., Quantitative Room and Space Requirements for the Proposed Marion Township High School Building, Marion Township, Franklin County, Ohio. The Bureau of Educational Research, College of The Ohio State University, 1950. (Mimeo.)

7. Lee, Ata, Space and Equipment for Homemaking Programmes: Washington: U.S. Government Printing Office, Office, 1950. (Federal Security Agency. U.S. Office of Education, Misc. No. 9)

8. McLeod, John W., "Storage Cabines Assemblies as Dividing Partitions," The American School and University, Twenty-first Edition, p. 223, 1949-50.

9. National Council on Schoolhouse Construction, Plant Guide Committee, Guide for Planning School Plants, Nashville, Tenn.; The Council, 1949.

10. Van Waynen, Martinus, "Mock-Ups in Science Teaching," The Science Teacher 18: 244-246, November, 1951.

11. Whitehead, Wills A., Suggests for Committees Concerned with Planning Elementary and Secondary School Facilities. The Bureau of Educational Research, College of Education, The Ohio State University, 1947 (Mimeo).

7

Basic Philosophy of Teaching Chemistry

1. The purpose of the study of chemistry in high school is not for mere fact mastery for its own sake, but through the use of facts and principles of chemistry and through the methods of approach to their mastery. to learn the art of reflective, criticial thinking in the solving of problematic situations. To accomplish this end certain principles should be observed:

(a) The pupil should learn clearly to define and state a problem, to recognize the various aspects entering into the problem, and open-mindedly to study the various phases of the total situation to discover possible solutions.

(b) The pupil should be trained to state conclusions tentatively, and to test such conclusions before arriving at a judgement.

(c) Emphasis should be placed upon the value of chemistry to the individual and to his social group, rather than upon organized subject matter as an end in itself.

(d) Pupils at any age level may be taught to use the methods of science, greater proficiency coming with added maturity and experience.

(e) Each pupil should learn what *things* are, what *words* mean, and what *concepts* teach.

(f) Each pupil should learn to think for himself, and at the same time to respect the opinions of others. For example, science and religion need not conflict if students will seek and accept the whole truth.

(g) Testing the results of thinking is essential. The slavish acceptance of a dictum on the basis of authority is unwaranted and unwise.

(h) Scientific curiosity is an important factor in critical thinking. All too often it is discouraged in pupils of elementary and secondary-school age.

(i) Subjective attitudes and emotional standards must be recognized for their true value, in which the study of chemistry can serve as a guide and a check.

2. To aid in accomplishing this end, the pupil should be enabled to develop the following qualities from a study of chemistry:

(a) An inquiring mind, with sufficient skill in speaking, reading, and calculating to enable him to participate intelligently in the life of the community.

(b) The ability to make accurate observations, unbiased by preconceived notions.

(c) A working knowledge of the essential contributions of chemistry to health.

(d) The ability to be a careful, discriminating user of scientific products.

(e) An adequate understanding and appreciation of the contributions of chemistry to the development and conservation of our human and material resources, and our industrial and commercial life.

(f) A broad mastery of the fundamental, basic factual material of the science to enable one to accomplish these results.

3. Constant emphasis upon the contribution of chemistry to the public welfare, and to humanitarian ends in general is calculated to aid in developing a sound philosophy of life.

4. One of the most pressing problems of our nation is the conservation of natural resources, to which end chemistry is prepared to contribute much.

5. A study of the contributions of chemistry to our economic, industrial, and social life gives abundant chance for integration of this area with the areas of the social studies and the humanities.

A Statement of Objectives for High-School Chemistry

1 To acquire a reasonable mastery of a chemical vocabulary, concepts and principles, an understanding of the laws and theories of chemistry, and an ability to apply such knowledge to appropriate situations in everyday life.

2. To arouse and stimulate the pupil's interest and curiosity in his environment, and to train him in scientific and creative thinking.

3. To help acquire an appreciation of the lives of great chemists and of their contributions to the welfare and progress of society.

4. To show the relationship of chemistry to industry, conservation of natural resources, agriculture, purchasing of economic goods, the home, medicine, health, and national defense.

5. To discover special aptitudes in students, and to give all possible necessary guidance and assistance to those who should consider the subject from a vocational or a professional point of view.

6. To acquire skills in laboratory techniques so that there may be developed the ability to solve laboratory problems with

skill, dispatch and resourcefulness, and to make accurate observations and exact statements.

Elements and Safeguards of Scientific Thinking*

The teacher of chemistry has an excellent opportunity, as well as a duty, to train pupils directly to think in the field of science. Teachers have believed that science was an organized body of subject matter which if taught, would result in the application and use of the scientific method by pupils in their daily living. John Dewey has pointed out that science is as much a method of thinking as it is organized subject matter.

Eckels (1) has said, "Too many teachers assume that exposure to a body of subject matter which has its foundation in, clear thinking automatically induces that process which led to its being. Recent surveys of business offices, coupled with the opinion of men like Nicholas Murray Butler, have shown that pupils are graduating from our high schools unable to think and formulate seasoned judgements. Strauss (2) has further shown that the pupil gains more in his ability to think scientifically in the tenth grade. As he proceeds on through high school and college, he actually loses his power critically to evaluate forked road situations presented to him. He may be learning to think in specific subject matter areas, but he is unable to transfer the scientific method and use it in other situations.

Dashiell (3) has indicated that, "Transfer of training may take place in some situations but not in others. Transfer lies not in subject, student's general memory power or attention or reasoning in general, but resides in those habits of responding in this or that way to that aspect or detail which is common to both situations." Transfer probably takes place when the

---

*Reprinted by permission from the Journal of Chemical Education. Vol. 18 (February 1941), pp. 92-94.
Critical evaluation made by Dr. Rufus Reed. Dr. Catherine Cassidy, and Workers Course, Western Reserve University, Cleveland, Ohio, Summer, 1940.

pupil has isolated a concept from its surrounding material and uses it when he sees a detail in the new situation is common with one in the old situation. If the pupil is loaded with subject matter so that he has no time to reason, and if he is not taught directly so to do, he will only memorize and selectively recall what he has learned.

A student who possesses meager, limited concepts associated with concrete, previous experiences, will not properly recall such experiences. Adequate concepts free from elements associated with them produces free and versatile thinking, such as the following.

*Observation of Metal Foils**

"I''d like to tell you about some observations I made over the week end. I was in the basement idling my time, and I ran across some lead. I had a melting pot and put some of the lead in it and melted it. This led me to think about the melting of different metals. (Association with element in past experience). As the foil off of different objects interests me (Interest element and pleasurable experiences eating candy bars), I obtained two different looking foils. (Purposeful observation). I took the fullest and heaviest small piece and melted it. It formed a substance like that that was formed on the melted lead, lead oxide or sulphide, I suppose. (Past experience—correct concepts). Next I took a lot of this foil and alas, it melted into a metal, lead, just as the other. (Cautious repetition of observations—controlled experimentation—seeing elements in common.) I had heard of lead-foil and tin-foil; these foils seemed amazing they were so thin. I wanted to find tin-foil and how different it was from lead-foil. (Spirit of discovery thoroughly awakened). This other foil I had was put into the pot and heated. The former was heated on a gas stove burner. This foil was bright and the heat bad no effect on it. I heated it longer, still nothing happened. I thought tin

*The material in quotation marks under this heading was contributed by Lewis Surfried, Normandy High School, St. Louis Missourf.

had a lower melting point than lead, but I checked myself to make sure. (Looked in books.)

"That eliminated tin. I looked in the table and saw that it must be aluminium which I had not thought about, and it had a much higher melting point. (A second tentative hypothesis).

"As I had an oxy-acetylene torch at my disposal, I heated the pot with this and obtained the desired temperature. (Controlled experimentation—allowing but one variable.) First I took a piece of sheet aluminium and melted it, and then I took the foil. Sure enough, they were the same. Well, after an interesting time, I can now tell the difference between lead and aluminum foil. However, I couldn't find a piece of tin-foil, but I shall do some experimenting with it. Could the substance formed on those molten metals of lead and aluminium be lead and aluminium oxide, respectively? (Additional problem.)"

In recalling past experiences the wide range of his experience was brought out. He certainly verified his inference or "hunch" about the aluminium foil experimentally, and he permitted only one variable, the increased heat of the oxy-acetylene torch.

An experiment some five months before had preceded the boy's work on foils. This experiment involved heating and noting the weight changes of samples of magnesium, zinc, and lead. The inerease in weight due to the formation of oxides was observed, except in the case of the lead. The lead showed no gain in weight, but the boy noticed that yellow sulphur had gathered on the sides of the crucible. He deduced that as fast as oxygen was taken on sulphur was given off, therefore, the sulphur was an impurity. Exceptions were given special attention in the case. In teaching for transfer of training, it is therefore well to present pairs of situations where one factor and an opposite is present. In this manner concepts were cleared.

Instructors can teach only what they know. I should be obvious that a teacher could not present the scientific method with safeguards in specifically organized pupil experiences unless he understood and used the scientific method himself. Chemistry teachers outside the laboratory too frequently become a prey to emotional appeals and prejudices of every sort. For this reason, the training of in-service teachers in the use of the scientific method is of primary importance.

Another teaching technique would be to present pairs of situations where one factor and an opposite would be present. The teacher might then define the element, indicate the class to which it belongs, the characteristics which distinguish it from other elements to which it is similar.

An example would be to test SO_2, CO_2 and CaO in distilled water with litmus. The action of any combination of metals in the electromotive series could be used and their reaction to HCl shown.

One caution, however, is that the teacher must be sure that the student has dissociated the concept from other elements occurring with it, otherwise hazy concepts occur, and any transfer to solving problems in new situations will be lost.

Failure on the part of the pupil to define a problem is due to several factors; the lack of adequate concepts; he has not the factors which will place the elements in a class clearly in mind; he lacks appropriate terminology to differentiate terms; illustrations, examples, or experimental activities were not presented in a forceful way; he has failed to set up evaluating criteria in his mind to check his results.

The failure to evaluate the competence of authorities studied, their techniques and materials, whether the work has been checked or whether the authority has an "ax to grind" may lead to blocked knowledge.

A boy brought a religious pamphlet to school, given him by his mother. The pamphlet indicated that the family would be poisoned if they are food cooked in aluminium utensils. A class discussion sown brought out that the man writing the pamphlet was not a competent authority.

In making a comparison, pupils will invariably cite facts which prove that two elements are in two separate classes. A comparison of CO and CO_2 would usually end in the pupil classifying the compounds as monoxide and dioxide. If the pupil's attention is directed to first identifying facts, and then preparing a list of points of likeness as well as points of difference, he would see that they were related on the basis of the amount of oxygen present. With the concept freed, at a later date, he might think of applying his knowledge to the stoking of furnaces.

An investigation by one pupil was the result of observing that a silver chloride precipitate dissolved in an excess of sodium chloride. He performed a number of experiments to test the solubility of silver chloride. He was careful to run controls and allow only one variable. After a number of experiments which resulted in blocked knowledge, he consulted a book. He was not satisfied with what the book said, so he decided on a course of action of his own. Pupils even at the college level are too prone to put down what the book says, rather than purpose-several explanations for their observations. The boy observed the light stayed on in a mass of silver chloride. One book said that silver chloride ionized when heated to the point of fusion, if carbon electrodes were used. The inference was that the silver chloride must ionize if it carried a current, but this high-school pupil wondered why the light in the circuit stayed on after the mass had cooled. He proposed several theories and verified one by amalgamating the silver with mercury. He drove the mercury off and found a bridge of silver carried the current, the silver having been reduced by the action of carbon on silver chloride. He selected the most probable consequences and followed it up. He was not satisfied with what the book said. The plan of action was not the easiest nor the most common method for studying the solubility of silver chloride.

Certain errors are common to the inductive type of thinking. An example would be, a pupil may assume that all foils contain tin, because the expression tin-foils is used. He might believe that assumption unless he checked it by testing candy bar wrappers. He might assume that all compounds ionize after testing a few inorganic compounds. His assumptions would be wrong because of a poor selection of experimental facts. He might make errors in observing wrong relationships. He might believe that the electron was as large as a small pea, just because a picture in a book had it drawn that way.

Sometimes appropriate or valid principles cannot be recalled. Again, the inference drawn may not be correct because more than one variable has been allowed during the process of experimentation. Other errors in drawing inferences may be; errors in premise, that is, assumed, supposed or proved propositions; ambiguities or double meaning in language; errors in emphasis; errors in inclusion and distribution; errors due to form of inference method; errors due to familiarity of person with subject matter; or mixing emotion and thought which results in grooved thinking and loss of flexibility. The human mind may escape reality and draw the wrong inferences by leaning on authority instead of personal investigation; thinking what exists is best; personifying impersonal forces, that is, thinking certain racial groups are responsible for misfortunes; attempting to influence peoples' thought by appeal to character of person rather than point of issue; appeal to authority, to force, to sentiment or attempt to throw the burden of proof on the challenger. This last is used by quacks, saying, "You do not know these influences do not exist."

Drawing inferences is one general way of thinking. A second method, as pointed out before, is to analyze situations and attempt to bring about a synthesis of the important elements in order to solve problems.

Imperfect analyses usually result when the person making the analyses fails to attend to essential elements or when his attention has been focused on the wrong elements because of repression, suppression or prejudice.

Pupils in studying for examinations frequently study what they think the teacher will ask for, instead of organizing their knowledge. The newer educational methods are moving away from emphasis on examinations and marks, and the criteria of evaluating a pupil performance will likely in the future be based on how constructively and tangibly the pupil uses his knowledge to solve problems presented in extra-curricular and socio-economic areas. The accumulation of conduct changes by jotting down the pupil's remarks or actions. The presentation of exhibits of pupil work are trends toward new evaluating techniques. The pupil may orally present to a lay audience evidences of his work in an area of knowledge. The value of such methods for training for future citizenship is enormous.

A third general type of thinking is that of making selections. Imperfect selections are usually the result of a narrow, restricted point of view based on habit, emotional inhibitions, or dominance of one's own purposes or wishes. One boy felt that it would not do to study for a chemistry examination, because study only confused him.

A fourth general method of thinking is to make associations. Our schools have been erected to strengthen associative bonds. A pupil in studying may forget to see point of likeness. He may not want to see a relationship, for it jars him out of his complacency. To keep the personality intact, pupils may place the blame for failures in another quarter. A girl may think she has the attributes to enter nursing, because some one has said that she looks or acts like a nurse. She may copy from others in high school, enter nurse's training, and be dismissed after a year for writing up her charts two hours in advance of the time temperatures are to be taken.

A fifth general method of thinking is generalization. Pupils cannot be expected to generalize until they have recognized, defined the problem, and have clear concepts which have been dissociated from other related factors in the setting. Imperfect generalizations result through not having sufficient data, using false authorities, reasoning by analogy and drawing hazy inferences. In reasoning by analogy the pupil may feel that because he had to collect bottles of oxygen under water, that all experiments involving the collection of gases must be done the same way. If a general impression of an individual colours his judgement about a matter in all respects, his thinking will be fuzzy.

The analysis of unknown metals or non-metals is helpful in getting the pupil to reason by the method of residues or the "process of elimination". In testing for an unknown metal, suppose the pupil saw an orange colour. He would have to recall what orange meant. If he made the correct association, he would think calcium, but he would have to eliminate other colours and words in his thinking until the correct association was made.

The development of Boyle's Law enables a resourceful chemistry teacher to show what is meant by reasoning by the method of variable factors occurring together, or as it is called in logic, the method of concomitant variation. If in a series of variable causes, one cause is associated with one effect, then one may be said to be dependent on the other. The temperature being constant, Boyle found that an increase in pressure to which a given volume of gas was subjected was accompanied by a corresponding decrease in the volume of that gas, and vice versa, that every decrease in pressure was accompanied by a corresponding increase in the gas volume. Boyle concluded that the change in volume of the gas was caused by the change in pressure.

The elements and safeguards of scientific thinking are the same for both inductive and deductive processes, only the sequences are changed. The inductive method is primarily used by the person who wants to make discoveries. This is perhaps the most difficult yet the most fruitful process for the teacher to use.

Through science clubs and classroom experiences, the pupil's attitudes must be formed and his emotions set off and focused before he will tackle the problem to be solved. It is doubtful if a pupil will take his place as a thinking citizen in the social order, unless he has an opportunity to practice acting on thinking in curricular and extra-curricular situations.

This means less emphasis on teacher activity and more emphasis on pupil activity. It means a teacher in the background, encouraging and building with only occasional appearances on the center of the stage. It means working out new evaluating techniques by creating organizations for the presentation of student papers, science fairs, etc.

A complete revision of our methods of teaching is in order. The subject matter of chemistry will become secondary to the scientific method. It will be clothed in more dignity as a tool, because it will be more useful and functional than ever before. Let us put away our prejudices and preconceived ideas and as true scientists help usher in this new age of chemistry.

Suggestions to Teachers

1. In the use of this course of study no teacher should feel that it is expected that he follow it slavishly. Each teacher should feel free to use his own devices and methods, using the state course of study as a guide to point up his thinking and to indicate the general direction in which he is to go.

2. The teacher should not assign technical words to be looked up by the pupil. The technical word lists are given in each unit for the purpose of pointing up the teacher's thinking, so as to make him conscious of the vocabulary difficulties that are sure to be encountered.

3. The teacher will feel free to use socialization in the development of any unit where, in his particular situation, this can profitably be done even though no suggested in the outline.

4. It is probably best, in the development of quantitative concepts, to begin fairly easily in the year to introduce simple exercises requiring quantitative thinking, and to continue to introduce such concepts from time to time, until the pupils have developed the ability to think quantitatively. Certain it is that we have no reason to expect pupils to be able to develop this ability overnight.

5. For the personal use of the teacher the following list of reference books is suggested:

   Frank, Joseph Otto, *The Teaching of High School Chemistry*, published by the author, Oshkosh, Wisconsin, 1932.

   Heiss, Elwood David, Obourn, E. S., and Hoffman, C. W., *Modern Methods and Materials for Teaching Science*, Macmillan, 1940.

   Noli, Victor Herbert, *The Teaching of Science tn Elementary and Secondary Schools*, Longmans, Green, 1939.

   Preston, Carleton Estey, *The High School Science Teacher and his Work*, McGraw-Hill, 1936.

   Downing, Elliot Rowland, *An Introduction to the Teaching of Science*, University of Chicago Press, 1934.

   Hunter, Francis D., *Science Teaching*, American, 1934.

   Woodring, Maxie Nave, Oakes, M. E., and Brown, H. E., *Enriched Teaching of Science in the High School*, Bureau of Publications, Columbia Uniyersity, 1928.

   Hawkes, Herbert Edwin, *Construction of Achievement Tests*, Houghton Mifflin, 1936.

6. A few magazines that will be found most useful in connection with this course are:

   The Journal of Chemical Education, $3.00, Easton, Pennsylvania.

The Science Digest, \$2.50, 631-643 St. Chair St., Chicago, Illinois.

The Science News Letter, \$5.00, 2101 Constitution Ave., Washington, D. C.

The Science Leaflet, \$1.50, 60 E. 42nd St., New York, N. Y.

The Science Teacher, \$2.00, 201 N. School St., Normal, Illinois.

Each of the above named magazines, except the first, will be found useful in other science courses, as well as in chemistry.

7. Since one of the outcomes of a good course in high-chemistry should be the ability on the part of the pupil to assemble and evaluate facts, and to arrange them into an orderly pattern so that definite conclusions may be reached, these conclusions should in each case be clearly stated. Then the conclusions reached should be tested by every means possible. It is felt that, if this process is kept in mind by the teacher in the development of each unit, the pupils should gradually learn to draw conclusions tentatively on the basis of all the facts available, and then to try by every means possible to verify the validity of the conclusions reached.

8. The pupil should br encouraged to prepare a day-by-day budget plan for his time, after he has read the assignment and the experiments. The pupil makes his own assignment, which is later checked by the teacher. In general it is suggested that experimental work provide supervised study on a given topic. Lectures may come either before or after written assignments, as the teacher may prefer. The pupil should be encouraged to lean to organize material, and to make plans for himself.

9. It is suggested that each teacher devise his own method of evaluation of the work done in each unit in

harmony with the best modern practice in testing. This might involve the use of standardizcd tests, informal observations, or the construction by the teacher of his own tests. In any case, it is probably best, for the most part, to sample widely from the material covered and to construct objective tests for the purpose of unit evalution. Brief short tests may be given every few days. These may be either essay or short answer type. Many teachers may want to use a pretest in order to measure achievements more accurately. If this is done, every effort should be made to construct two substantially equivalent forms of each unit test. In this course of study no specific test items are suggested in the unit outline.

List of Illustrative Units in Chemistry

UNIT I—The Nature of Matter Challenges Man

UNIT II—The Structure of Matter

UNIT III—Types of Chemical Change

UNIT IV—Water, Solutions, and Near-Solutions

UNIT V—The Metals and Their Compounds

UNIT VI—Useful Nonmetals

UNIT VII—Carbon and Its Compounds.

NOTES

1. Eckels, C. F., "Clear Thinking Through Use of Subject Matter Materials," *California Journal of Secondary Educatian*, Vol. 11, 1936, pp. 56-57.
2. Strauss, Samuel, "Some Results from Tests of Scientific Thinking," Masters Thesis, University of Chicago, 1931.
3. Dashiell, J. F., National Society for the Study of Education, 37th Yearbook, Part 2, p. 382.
4. "Materials Prepared by Participants in Biological and Physical

Science Evaluation Group Summer Workshop," University of Chicago, Chicago, Illinois, 1939.

5. Pruitt, C. M., "An Analysis, Evaluation, and Synthesis of Subject Matter, Concepts, and Generalizations in Chemistry," Ph.D. Thesis, Teachers College, Columbia University, New York, 1935.

6. Symonds, P. M., *Education and the Psychology of Thinking*, McGraw Hill, 1936.

8

Unit I
The Nature of Matter Challenges Man

I. Specific Objectives

A. To develop an awareness of the effect chemical discoveries have had upon the progress of man.

B. To show that matter undergoes changes, the factors that cause chemical change, how to control these changes, and the effect that these changes have upon man and his surroundings.

C. To help pupils to gain a realization that nature follows an orderly course, as shown by certain laws concerning matter and energy; namely, the Law of Definite Proportion, the Law of Conservation of Matter, and the Law of Conservation of Energy.

D. To enable pupils to become familiar with the fact that the speed and coherence of atoms and molecules make up the physical state of matter.

E. To show that pure substances may be identified by their specific physical and chemical properties and that these properties are unchanged in the production of mixtures but changed in the production of compounds.

F. To assist pupils to acquire a knowledge of the preparation, properties, and uses of oxygen and hydrogen and of their significance to man.

II. Vocabulary

Physical property	electron	allotropic form
chemical property	proton	catalysis
element	molecule	electrolysis
compound	dissolve	oxidizing agent
mixture	residue	combine
physical change	law	chemical activity
chemical change	energy	theory
matter	combustion	electrochemical theory
state of matter	oxidation	electromotive series
composition atom	reduction	replacement

III. Subject Content

A. Man has always been Concerned with the Changes in Matter.

1. These changes touch our daily lives.
2. The Ancients viewed changes with suspicion.
3. The Alchemists wished to understand and control changes.
4. Discoveries and inventions have resulted from man's desire to answer the questions "Why?" and "How?"

B. Some Changes are Complex, some Simple

1. Changes in properties foretell changes in composition.

(a) Matter is recognized by its properties, that is, its physical characteristics and its chemical behaviour.

(b) Changes may be of a physical or a chemical nature.

2. Nature recognizes three types of material.

(a) Elements are the "building blocks" or units.

1. Symbols are used to express ideas.
2. The distribution and importance of elements vary.

(b) Compounds are the structures formed by the union of elements.

1. The structure follows a definite plan. (The Law of Definite Proportion; Composition)
2. Chemistry illustrates the principle of perfect balance.
3. World equations and formulas are interesting chemical tools.

(c) Mixtures are produced by the proximity of materials, not by a chemical union, and no definite proportion is required.

3. Most changes can be controlled.

(a) There are two types of changes to be considered, Chemical change differs from physical change.

1. In a chemical change the original substance loses its properties, and a different substance having different properties results.
2. Physical changes are changes in form, not in composition.

(b) Changes in matter are accompanied by changes in energy.

1. Matter undergoes change but does not lose or gain weight in so doing. (The Law of Conservation of Matter; Mass)
2. Heat energy transformed into molecular energy shows that matter may exist in any one of three states, gaseous, liquid, and solid. (Kinetic Molecular Theory)
3. Energy changes take place but no energy is lost or gained. (The Law of Conservation of Energy)

(c) Many factors are involved in chemical change.

1. Heat causes many changes (to accelerate change).
2. Catalysts also change the speed of a reaction.
3. Light and electricity are other forms of energy which affect chemical action.
4. Bacteria have been found to play a prominent part in certain types of chemical change.

(d) Matter and energy changes can be used to promote the welfare of man.

1. Chemical changes can be used to produce heat, light, and electricity as required.
2. Energy can be used to transform raw materials into desired chemicals, or useful substances.

C. The Element Oxygen can be Produced Through Chemical Change and can Participate in Changes.

1 Oxygen is more abundant in the earth's crust than any other element. It is found in the free state, in mixtures, and in many compounds.

2. Oxygen can be prepared by mechanical and chemical processes.

(a) Air can be liquefied and boiled (commercial method).

(b) Water can be decomposed (explanation of electrolysis of water).

(c) Sand contains oxygen in abundance but oxygen cannot easily release its hold on its companion silicon; therefore, compounds whose elements have less affinity for each other must be chosen.

1. The historical method used mercuric oxide.
2. The laboratory source is usually heating potassium chlorate.
   (The above method works. Try adding a little manganese dioxide; it's like adding fuel to a flame—catalysis.)

3. The properties of oxygen make it a boon and a bother to mankind.

(a, Physically oxygen has no outstanding characteristies.

(b) Chemically oxygen is a problem.

1. It is an allotropic element. The characteristics of oxygen bear investigation.
2. It combines with many different materials. It is very active element.
   (These reactions produce heat—exothermic reactions.)
3. The union of oxygen with an element is called oxidation; its products, oxides. There are two typcs of oxidation, slow and rapid. (The latter is accompanied by heat and light and is known as combustion.)

(c) Oxidation may be a process helpful to man.

1. Under control it helps heat our buildings. (To control it three factors must be considered: presence of fuel, attainment of kindling point, and the proper surrounding medium.)
2. It furnishes heat from oxidized foodstuffs digested in the body.
3. It helps destroy bacteria in water, sewage, etc.
4. It helps eliminate waste by promoting decay.

(d) Oxidation may be a nuisance to man.

1. Spontaneous combustion causes great losses.
2. If foodstuffs are not protected from the action of oxygen they spoil.
3. Houses unprotected from oxygen decay.
4. Fires, uncontrolled by failure to remove one of the factors necessary for combustion, produce great loss of life and property.
5. Dust explosions are numerous and destructive; they may be prevented.
6. Metals corrode due to the action of oxygen; water often acts here as a catalyst.

D. Hydrogen, Copartner of Oxygen

1. Hydrogen is too "restless" to remain alone.
   Since hydrogen "loves" company it is found in all sorts of combination. Some combinations are water, acids, bases, petroleum products, etc.

2. Due to its activity it must literally be "pushed" out of combination.

   (a) Water releases all its hydrogen under a drastic process such as electrolysis.

(b) The activity of metals may be charted by testing their ability to "push" hydrogen out of combination. (Electromotive Series)

1. Water usually loses only one half of its hydrogen content when under metallie attack. All metals do not release hydrogen from water.
2. Most acids liberate their hydrogen in the presence of active metals.

3. Certain properties of hydrogen make it a useful element.

(a) Its density gives rise to a limited use in aircraft.

(b) Its chemical activity makes it valuable industrially.

1. Food products are modified by its use. (Hydro-genation)
2. Metals are extracted from their ores due to its reducing power.
3. It is used in cutting and welding metals.
4. The petroleum industry has used it for increasing output.

4. Other properties of hydrogen lower its value and even prove it to be dangerous.

(a) Helium is replacing hydrogen in aircraft.

(b) Diffusion brings an interesting glimpse into the world of moving molecules.

IV. Learning Activities

A. Teaching Suggestions

1. Place a copy of "What has Chemistry to do with me?" Found in the introduction of *Chemistry in*

Industry, Volume I, by Howe, at each place in the classroom and the class is off to a running start.

2. Display pictures of early scientists to interest pupils in scientists as personalities.

3. Define the broad field of study, point out the objectives for the unit, give examples of areas to which the subject contributes, i.e., medicine, agriculture, industry, etc. Survey the vocational interests of the group.

4. Display a group of elements, such as zinc, copper, iron, sulphur, carbon, etc., and from these develop an understanding of elements and their properties.

5. Heat iodine in a flask to illustrate changes in physical properties, etc.

6. Heat a piece of paraffin in a test tube to bring about a discussion of physical change and states of matter.

7. Ignite a mixture of zinc dust and sulphur on an asbestos square and use this to start a discussion of chemical changes, energy changes, and the formation of compounds. Introduce word equations.

8. Display powdered lead nitrate and powdered potassium iodine; then grind them together in a mortar to show chemical change involving compounds. Use word equations.

9. If a bit of phosphorous (red or yellow) or sulphur is ignited and thrust into a bottle (oxygen) which has been standing covered on the desk, a discussion of oxygen will usually developed spoptaneously.

10. To introduce the subjects of decomposition and catalysis, heat a small quantity of potassium chlorate, keeping track of the time required to produce sufficient oxygen to ignite a splint. Then heat a similar amount of manganese dioxide over

the same flame, for the same period of time, testing time and again for oxygen. Finally mix similar quantities of the two original materials and heat this mixture over the same flame, again timing the process. The question of the similarity of amount of oxygen produced in each case might well be considered here and the Law of Definite Proportions (composition) emphasized.

11. Have a tall cylinder containing mossy zinc, dilute hydro-chloric acid, and moth balls on the desk when the class assembles. A discussion of the preparation and properties of hydrogen should develop readily.

12. Pass *pure* hydrogen through a drying agent and over some heated black copper oxide.

13. Encourage pupils to look for material pertinent to the unit's work by making use of such material frequently.

14. Assign pupil reports as nearly as possible on the basis of varied interests, as determined by a questionnaire or volunteer choice.

Suggested Reports

The Iatrochemist

The Elemments of the Alchemists

Alchemy and the Development of Chemistry

The Phlogiston Theory

Forerunners of Chemistry

Catalysts in Industry

Roger Bacon, the Scientist

Suggested References for Preceding Material

Beery, *Stuff*

Darrow, *Story of Chemistry*

Clarke, *Marvels of Modern Chemistry*
Findlay, *Chemistry in the Service of Man*
Foster, *The Romance of Chemistry*
French, *The Drama of Chemistry*
Howe, *Cnemistry in Industry*
Jaffe, *Crucibles*
Moore, *History of Chemistry*
Sadtler, *Chemistry of Familiar Things*
Weeks, *Discovery of the Elements*
Wilson, *Great Men of Science*

B. Pupil Activities

1. Make a list of happenings in your daily life directly attributable to chemistry or a list of scientific achievements which have contributed to your health and happiness.
2. Make a short report on the contributions of certain early scientists, such as Priestley, Lavoisier, Cavendish, etc. Point out the difficulties they encountered.
3. Make a list of changes occurring about you and classify them under these headings: physical change and chemical change.
4. Make a list of materials used about the home and classify them under these headings: elements, compounds, and mixtures. Give the form or state of matter in which they normally exist.
5. Report the sources of any reading materials you discover which pertain to the subject under discussion, or keep a bibliography.
6. Bring to class any newspaper clippings which deal with the subject you are studying and add them to the bulletin board collection. Your name and the date should be on each article. An individual

mounting greatly enhances the appearance of the bulletin board.

C. Experimentation

Note: Pupils should be encouraged to perform as many experiments as possible within each field. Some possible experiments are suggested (directions for which are obtainable in any good laboratory manual). In addition, a few experiments are outlined to assist the pupil in further investigation where the desire is present, the time available, and the conditions make such possible.

1. Changes in matter

(a) Physical and chemical changes
(b) Weight and chemical change

1. Heating a known weight of a metal
2. Action of silver nitrate and sodium chloride solutions upon each other; weighed amounts to be used
Mixing weighed amounts of barium chloride and sodium sulphate

(c) Elements, compounds, and mixtures
(d) Production of heat, light, and electricity by means of a chemical change

1. Mix a base and an acid.
2. Burn a match, sulphur, etc.
3. Show electric current produced by means of zinc and copper strip in dilute sulphur acid solution.

2. Oxygen

(a) Preparation and properties of oxygen
(b) Oxidation

1. Slow oxidation
Coat the inside of a moistened test tube with iron filings. Place the test tube

mouth downward in a beaker of water and set aside until the following day.

2. Rapid oxidation

Burn a candle, heat a piece of copper metal, or heat zinc with a blowpipe.

3. Hydrogen

(a) Action of acids on metals
(b) Preparation and properties of hydrogen

D. *Optional Work for Individual Differences*

The teacher is encouraged to make provisions for individual differences. Some pupils will be able to work faster and more effectively than others. The ones who are able to work faster and do more work should not be penalized by having the more retarded pupils set the pace for them. Neither should the slower pupils be handicapped by being placed under the necessity of maintaining a pace that is inconsistent with their ability. There are many optional experiments, exercises, and studies which pupils of greater ability might be encouraged to do that would challenge them and develop their ability. Reference to good laboratory manuals will suggest such optional experiments.

In addition to optional experiments which might be performed and reports made to the class, the pupils might be encouraged to make written reports on studies made on some subject that is pertinent. A few such topics are as follows: "The Problem of Fire Hazards," "The Life of Priestly and of Scheele," "The Life of Cavendish." "Boyle and *The Skeptical Chymist*," "Langmuir and the Automic Hydrogen Torch," "Lavoisier and his Work," etc.

Planned conferences may be held with a member of the fire department, the local water works, the medical, legal, or dental profession of the community to bring out the part which chemistry plays in the daily work of each of these occupations. These conferences are especially valuable if the pupil is considering chemistry from a possible vocational point of view.

NOTES*

1. Biddle and Bush, *Dynamic Chemistry*, Rand McNally, 1937, pp. 3-69.
2. Black and Conant, *New Practical Chemistry*, Macmillan, 1936, pp. 1-60.
3. Brauer, *Chemistry and its Wonders*, American, 1938, pp. 1-85.
4. Brownlee et al., *Chemistry in use*, Allyn and Bacon, 1939, pp. 1-22.
5. Brownlee et al., *First Principles of Chemistry*, Allyn and Bacon, 1940, pp. 1-45.
6. Bruce, *High School Chemistry*, World Book Company, 1938, pp. 1-40.
7. Dull, *Modern Chemistry*,Henry Holt, 1940, pp. 3-39.
8. Gray, Sandifur, and Hannah, *Fundamental Chemistry*, Houghton Mifflin, 1938, pp. 1-21.
9. Greer and Bennett, *Chemistry*, Allyn and Bacon, 1939, pp. 1-37; 81-152.
10. Hopkins et al., *Chemistry and You*, Lyons and Carnahan, 1939, pp. 1-87.
11. Horton, *Modern Everydcy Chemistry*, Heath, 1937, pp. 1-82.
12. Howard, *Units in Chemistry*, Henry Holt, 1938, pp. 3-70.
13. Jaffe, *New World Chemistry*, Silver Burdett, 1940, pp. 1-66.
14. Kruh, Carleton, and Carpenter, *Modern-Life Chemistry*, Lippincott, 1937, pp. 3-64.
15. Lemkin, *Visualized Chemistry*, Oxford Book Co., 1938, pp. 1-28.
16. McPherson, Henderson, and Fowler, *Chemistry at Work*, Ginn, 1938, pp. 3-113.

* More complete bibliographical data on the references listed in each unit is found on page 447.

CHEMISTRY

TEACHER EVALUATION AND SUGGESTIONS FOR IMPROVEMENT

Unit ....

(*Number*) (*Title*)

Note to Teacher: As a result of your experiences in teaching this unit, please make suggestions for improvement and send to Lloyd W. King, State Superintendent of Public Schools. Use an additional sheet if necessary.

Specific Objectives

Subject Content

Learning Activities

9

Unit II
The Structure of Matter

I. Specific Objectives

A. To show how man has developed the modern concept of the nature of atoms, and the place that electrons, protons, and neutrons have in such structure.

B. To develop the concept of the structure of atoms as a means of helping to make clear such concepts as positive and negative valence, polar valence, non-polar or co-valence in terms of sharing and transfer of electrons.

C. To show how, by proper understanding of these facts, one can come to a better understanding of such matters as the order of activity of metals, and the differences between metals and nonmetals.

D. By means of a study of the Periodic Law, to enable pupils to come to a better understanding of the classification of elements into similar groups, and, by means of this relationship, to prediet the properties of elements that have not been discovered, or are not familiar to the pupil.

E. To show how, by a study of atomic numbers and electronic shell structure, the periodic classification of the elements is confirmed and enriched.

II. Vocabulary

diffusion	atomic number	vacuum tube
miscible	cathode rays	amphoteric
kinetic	positive rays	co-valence
neutron	alpha particle	polar valence
atomic weight	beta particle	valence
molecular weight	gamma rays	

III Subject Content

A. The atom is the ultimate particle that would result if a mass of an element were indefinitely divided until the ultimate particle was reached which would possess the properties of the element.

B. A molecule is the ultimate particle that would result if a mass of a compound were indefinitely divived until an ultimate particle was reached that would possess the properties of the compound. How would a molecule differ from an atom?

C. Trace the development of the concept of the atom, beginning with the ideas of the old Greek philosophers, and tracing it down through the theory advanced by John Dalton, and down to the present concept.

D. How did the discovery of radioactivity modify our concept of the atom? In what respect did it necessitate a restatement of the theory?

E. What evidence do we have of the existence of protons? What are they, and how do they differ from electrons?

F. How did scientists come to know of the existence of electrons? What are they, and what part do they play in the atomic structure?

G. What are neutrons, and where are they found in the atoms?

H. How are electrons, protons, and neutrons arranged in the atoms? (A few of the simpler atoms should be taken for examples, and their atomic stuctures worked out. Show the isotopes of each element selected for illustration).

I. Why is it said that the physical properties of an element are determined by the structure of the nucleus?

J. How do the electrons in the outer shell determine the chemical properties of an element?

K. A typical metal is an element which is capable of transferring one or more electrons from the outer electronic shell. The activity of a metal is measured by the readiness with which it makes this transfer.

L. A typical nonmetal is an element that tends to accept one or more electrons transferred from a metal to fill its outer electronic shell.

M. How can one determine positive and negative valence from the structure of an atom?

N. What happens when an element such as carbon unites with oxygen? How does this differ from the polar valence? How does this illustrate co-valence?

G. How did Mendelejeff discover and state the Periodic Law? Show how this law may be used in predicting the properties of an element. Of what practical value has this law been?

P. How did Moseley discover the atomic numbers, and how did this discovery lead to a restatement of the Periodic Law? Develop rather fully Moseley's great contribution and its value to modern chemistry.

IV. Learning Activities

A. Pupil Activities (Reports which may be done by committees or by individuals).

1. The ancient Greek ideas of atoms, molecules, and elements.
2. The life and work of John Dalton.
3. The life and work of H.G.J. Moseley.
4. The life and work of Mendelejeff.
5. The discovery of radioactivity, and how it modified our concept of the atom.
6. The modern atomic theory.

Lectures by the teacher and class discussion should supplement the work of pupils in developing the modern concept of the atom historically

B. Teacher and Class Activity

1. The teacher should take pains to lead the discussion, by lecture or by directing class discussion, so that the class comes to a clear conception of how our modern atomic theory has come to be, and what practical uses we make of the concent of the atom.
2. In similar manner the distinction between atoms and molecules should be developed.
3. By means of lectures, largely, the teacher may show how our ideas of the structure of the atoms as consisting of electrons, protons, and neutrons has come to be, and the manner in which this idea of atomic structure helps to explain valence, chemical activity, metallic and non-metallic character. This should lead to a clear conception of polar valence, positive, and negative valence.
4. The teacher should lead the class to see how the Periodic Law was developed, how it helped to promote the discovery of new elements, and how it aids man to understand the properties of little-known elements. Drill on this will soon show the

Periodic Law to be a very useful instrument in the study of chemistry.

5. The teacher should begin with the simple Crooks tube then the X-ray tube, show how Moseley, in measuring the influence of the elements on the wave length of X-rays, determined the atomic numbers, and how this discovery led to a restatement of the Periodic Law.

C. *Optional Pupil Activities*

1. Write out a clear statement of the contribution of Mendelejeff to the development of modern chemistry.

2. State the Periodic Law in terms in the most accurate information we possess.

3. Work out a plan of making a model of one or more atoms, showing the structure.

4. Brief report topics can be assigned from the subject content covered. Co-ordinate valence might be a good report topic. The relationship of various types of radiation to medicine and dental practice would also be a good topic.

NOTES

1. Biddle and Bush. *Dynamic Chemistry*, Rand McNally. pp. 107-151.

2. Black and Conant, *New Practical Chemistry*, Macmillan, 1936, pp. 173-185.

3. Brauer, *Chemistry and Its Wonders*, American, 1938. pp. 381-415,

4. Brownlee et al., *Chemistry in Use*, Allyn and Bacon, 1939, pp. 70-103.

5. Brownlee et al., *First Principles of Chemistry*, Allyn and Bacon, 1940, pp. 65-79; 128-150; 371-389.

6. Bruce, *High School Chemistry*, World Book Company, 1938, pp. 57-63; 353-365; 81-91.
7. Dull, *Modern Chemistry*, Henry Holt, 1940, pp. 26-40.
8. Gray, Sandifur, and Hannah *Fundamentals of Chemistry*, Houghton Mifflin. 1938, pp. 105-120; 465-476.
9. Hopkins et al., *Chemistry and You*, Lyons and Carnahan, 1939, pp. 139-176.
10. Howard, *Units in Chemistry*, Henry Holt, 1938, pp. 175-240.
11. Jaffe, *New World in Chemistry*, Silver Burdett, 1940, pp. 82-90; 168-195.
12. Kruth, Carleton, and Carpenter, *Modern-Life Chemistry*, Lippincott, 1937, pp. 65-135; 350-366.
13. McPherson, Henderson, and Fowler, *Chemistry at Work*, Ginn, 1938, pp. 162-178; 198-209.
14. Wilson and Mullins, *Applied Chemistry*, Henry Holt. 1939, pp. 205-232.

CHEMISTRY

TEACHER EVALUATION AND SUGGESTIONS FOR IMPROVEMENT

Unit.. -..............................
(*Number*) (*Tittle*)

Note to Teacher: As a result of your experiences in teaching this unit, please make suggestions for improvement and send to Lloyd W. King, State Superintendent of Public Schools. Use an additional sheet if necessary.

Specifie Objectives

Subject Content

Learning Activities

10

Unit III
Types of Chemical Change

I. Specific Objectives

A. To acquaint pupils with the concepts of atomic weights and molecular weights.

B. To enable pupils to understand and use symbols, formulas, and equations, and to learn the four important types of chemical change: combination, decomposition, displacement, and double decomposition.

C. To assist pupils in learning the symbol and valence of each of the common elements and radicals.

D. To help pupils in learning to name acids, bases, salts, and oxides, and to recognize the names from the formulas and vice versa.

E. To develop an understanding of the meaning of oxidation and reduction.

F. To enable pupils to learn the use of proportion in solving simple chemical problems.

II. Vocabulary

equation	divalent	alkalies
valence	trivalent	Berzelius
ions	balancing	Avogadro
synthesis	reduction	Gay-Lussac
analysis	oxidation	Boyle
replacement	borrowing	Charles
decomposition	lending	hypothesis
reaction	sharing	Dalton
neutralization	oxides	theory
hydrolysis	hydrides	equilibrium
electrolysis	subscripts	dynamic
electrode	coefficients	conservation of mass
anion	radicals	percentage composition
cation	anhydrides	
acids	reversible	atomic weight
bases	precipitate	molecular weights
salts	nonionivation	heat of formation
univalent	caustic	

III. Subject Content and Learning Activities

A. Chemical Symbols as we hare them today were Suggested by Berzelius.

Each element has a symbol which stands for one atom of that element.

1. Each element has a positive and negative valence which it, theoretically, may exercise. The only exception to this rule is the fact that the valence of the inert gases is zero.
2. Typical metals tend always to exhibit positive valence. while typical nonmetals tend to exhibit negative valence.
   (a) Metals differ in their readiness to give up their valence electrons. This leads to the order of activity of the metals.

(b) A similar difference is noted in the readiness with which nonmetals take on electrons to fill their valence shells.

3. The symbol and valence of each of the following named elements should be learned (preferably by usage):

aluminium	hydrogen	phosphorous
antimony	iodine	potassium
barium	iron	silicon
bromine	lead	silver
calcium	magnesium	sodium
carbon	manganese	sulphur
chlorine	mercury	tin
copper	oxygen	zinc
fluorine		

4. Radicals are groups of atoms combined and acting as a unit in chemical changes. Thase radical have each a definite formula and a definite valence. The formula and valence of each of the following named radicals should be learned (preferably by usage):

ammonium	nitrate	sulphate
acetate	nitrite	sulphite
bicarbonate	phosphate	sulphide
chlorate	phosphite	carbonate
hydroxide	silicate	chloride

5. Atoms combine in definite proportion to form compounds. Each compound has a definite formula which is invariable.

(a) In writing a formula more than one atom of a given element is represented by a subscript.

(b) The metallic element in a compound is always written first in the formula.

(c) Some gaseous elements, such as oxygen and hydrogen, are said to be diatomic. That is, they have two atoms to a molecule in the gaseous state. This idea should be developed.

(d) In writing a formula, the sum of the positive valence of the metallic ion and the negative valence of the non-metallic ion equals zero. This determines the number of each ion that is needed in a formula.

(e) Each formula represents a defiuite weight. It would be well for each pupil to prepare a list of the formulas of a number of common compounds.

6. The formula of each of the common acids and bases should be thoroughly learned. The following is a suggestive list (learned preferably by usage):

Acids	*Bases or Hydroxides*
hydrochloric acid	ammonium hydroxide
hydrobromic acid	sodium hydroxide
hydroiodic acid	potassium hydroxide
hydrofluoric acid	barium hydroxide
hydrosulphuric acid	calcium hydroxide
sulphuric acid	magnesium hydroxide
sulfurous acid	aluminium hydroxide
phosphoric acid	ferric hydroxide
phosphorous acid	ferrous hydroxide
carbonic acid	
nitric acid	
nitrous acid	
chloric acid	
perchloric acid	
chlorous acid	
hypochlorous acid	

7. The pupil should learn to calculate the molecular weight of a compound from its formula.

(a) Each element has a definite atomic weight, which is a ratio based upon the atomic weight of oxygen arbitarily taken as 16.

(b) The sum of the atomic weights of all the atoms in a molecule is its molecular weight which is the ratio of the weight of that molecule to the assumed weight of an atom of oxygen.

(c) Since the atomic weight of an element or the molecular weight of a compound is a ratio, it may be expressed in any convenient unit of weight, such as grams, pounds, etc. Thus we have gram atomic weights, or gram atoms, and gram molecular weight, gram molecules, or moles.

(d) The pupil should learn to calculate the percentage composition of a compound from its formula. A little drill on this will fix this point.

(e) If the gram molecular weight of any gas be measured at zero Centigrade and 760 mm. pressure, the volume will be about 22.4 liters. This is called the Gram Molecular Volume. This has many uses in chemical calculations. Among these uses is the ready calculation of the weight of a liter of any gas, if its molecular weight is known.

8. The Law of Definite Composition controls the matter of the formula for any compound.

B. *An Equation is a Graphic Representation of what Takes Place in a Chemical Change, and is Both a Qualitative and a Quantitative Expression of that Change*

1. A simple type of chemical change is combination. This may be illustrated by the chemical union of zinc and sulphur, or of oxygen and hydrogen.

2. Another type of chemical change is decomposition. This may be represented by the decomposition of mercuric oxide.
3. A third type of chemical change is displacement. This may be represented by the interaction of zinc and sulphuric acid.
4. A fourth type of chemical change is double decomposition. This may be represented by the interaction of hydrochloric acid and sodium hydroxide.
5. In writing an equation the following suggestions are given for the consideration of the pupil:

   (a) Determine what substance or substances enter into the chemical reaction, and write the formula for each such substance on the left side of the equation. The reacting substances may be written in any order.

   (b) Determine what product or products are formed, and write the formula of each such product on the right side of the equation.

   (c) Balance the equation. In doing this it is generally well to begin with the formula that is most complex in order to determine how many times each ion must be taken, since this determines the number of that ion which must be accounted for on the other side of the equation. This usually gives a definite idea as to how many molecules of each substance must be taken, and how many molecules of each product are formed. Coefficients are used to indicate the number of molecules of each compound in the equation. If a gaseous substance is found on either side of the balanced equation the coefficient also indicates the number of volumes of that gas. It is well, after the equation is thought to be balanced, to check each ion to see that the same number of each ion is found on each side of the equation and that each formula is correct.

(d) In most chemical reactions the radicals, such as the sulphate, nitrate, hydroxyl, or phosphate radicals, remain unchanged.

6. The pupil is to be asked to classify each of the following indicated reactions, and complete the equation:

   (a) Zinc and hydrochloric acid

   (b) The reduction of ferric oxide by carbon

   (c) The reaction of oxygen and hydrogen sulphide

   (d) Zinc and silver nitrate

   (e) Iron and sulphur

   (f) Silver nitrate and sodium chloride

   (g) Aluminum chloride and ammonium hydroxide

7. It is suggested that the teacher have the pupil write the equations for each chemical change observed in the laboratory work as an activity exercise.

C. *Oxidation and Reduction*

1. Oxidation involves the increase of the positive valence of an element, which is brought about by the loss of one or more electrons from the valence shell. This is illustrated by the following equations:

$$FeCl_2+Cl^0=FeCl_3 \text{ or } 2H_2+O_2=2H_2O$$

For example, the ferrous ion yielded one electron to the chlorine, to form an additional chloride ion. Thus the iron was oxidized, because it lost an electron, and the chlorine was reduced because it gained an electron. In the same way, the hydrozen lost an electron to the oxygen.

2. It would be well to have the pupils study several simple combination and displacement reactions in order to discover the electronic changes that occur.

NOTES

1. Black and Contant, *New Practical Chemistry*, Macmillan, 1936, pp. 101-116.
2. Brownlee et al., *Chemistry in Use*, Allyn and Bacon, 1939, pp. 291-309.
3. Bruce, *High School Chemistry*, World Book Company, 1938, pp. 131-142.
4. Dull, *Modern Chemistry*, Henry Holt, 1940, pp. 33-39.
5. Gray, Sandifur, and Hannah, *Fundamentals Chemistry*, Houghton Mifflin, 1938, pp. 121-137.
6. Hopkins et al., *Chemistry and You*, Lyons and Carnahan, 1939, pp. 229-261.
7. Jaffe, *New World of Chemistry*, Silver Burdett, 1940, pp. 124-134.
8. Kruh, Carleton, and Carpenter, *Modern-Life Chemistry*, Lippincott, 1937, pp. 239-250.

CHEMISTRY

TEACHER EVALUATION AND SUGGESTIONS FOR IMPROVEMENT

Unit...,............
(*Number*) (*Title*)

Note to Teacher: As a result of your experiences in teaching this unit, please make suggestions for improvement and send to Lloyd W. King, State Superintendent of Public Schools. Use an additional sheet if necessary.

Specific Objectives

Subject Content

Learning Activities

11

Unit IV
Water, Solutions and Near-Solutions
(Acids Bases and Salts)

I. Specific Objectives

A. To show how our daily life is affected by water.

B. To study the properties of the various kinds of solutions and their applications to the study of chemical change, .as well as the use of normal and molar solutions.

C. To study the various ways in which solution for electrolytes differ from those of non-electrolytes.

D. To develop the various ways in which the theory of ionizatian helps to explain various phenomena in the fields of chemistry, biology, medicine, and industry. (Among these phenomena should be considered the cause of conductivity of an electrolytic current through a solution of an electrolyte, the explanation of the mechanisms of chemical reactions, and the reasons for the fact that solutions of electrolytes have an abnormally low freezing point, an abnormally high boiling point, and an abnormally high osmotic pressure).

E. To study the nature and properties of simple colloidal systems, and some of their applications in nature, in industry, and in the home.

II. Vocabulary

non-electrolytes	burette	hydrogen ion concentration
electrolytes	end-point	hydroxyl ion concentration
dissociation	deliquescent	normal solution
neutrality	precipitation	molal solutions
titration	volatile	ions
hypothesis	concentrated	efflorescent
dilute	coagulation	mole
conductivity	adsorption	molar solution
super-saturated		

III. Subject Content

A. The Occurrence, Properties, and Uses of Water

1. Natural sources of water

(a) Wells

1. Shallow wells, dangers of, and safeguards against contamination
2. Deep wells

(b) Springs

1. Problems of maintaining purity of water,
2. Study some of Missouri's large springs, such as Big Spring, Greer Spring, and Bennett Spring.

(c) Streams

Problems of pollution

(d) Ponds and lakes

Possible contamination, and means of control of such dangers

2. Purification of municipal water supplies.
3. The important physical properties of water
   (a) It reaches its maximum density at about 4C°, at which point its density is one, and one cubic centimeter of it weighs 1 gram.
   (b) Vapour pressure in relation to temperature; boiling point of water under various conditions of pressure.
   (c) The universal solvent. Study several typical examples of the use of water as a solvent.
   (d) Heat of vapourization of water.
   (e) Heat of fusion of ice.
4. Hardness of water.
   (a) Temporary hardness, its causes and methods of removal.
   (b) Permanent hardness, its causes and methods of removal.
   (c) Industrial problems involving the use of hard water.
5. The chemical properties of water.
   (a) Its analysis by electrolysis.
   (b) Its reaction with nonmetal oxides, forming acids.
   (c) It reacts with oxides of the more active metals, forming balsic hydroxides.
   (d) Many crystalline substances have water chemically combined forming hydrates.
6. Heavy water.

B. Solutions

1. In a solution the solute (the material dissolved) exists as single molecules dispersed through the solvent.

2. Factors affecting the solubility of a solute in a solvent.

   (a) The nature of the solvent.
   (b) The fineness of division of the solute.
   (c) The temperature.
   (d) The solubility of the solute in the solvent.
   (e) Whether the system is stirred or agitated during solution of the solute in the solvent.

3. Saturated solutions, and how they may be formed.
4. Super-saturated solutions; methods of preparing; their instability.
5. Molar solutions contain one mole of solute per liter of solution.
6. A molal solution contains one mole of solute per 1000 grams of solvent.
7. A normal solution contain one equivalent weight of the solute per litre of solution.
8. The effect of a solute on freezing point and boiling point of a solvent. One mole of a non-electrolyte per litre of solution gives a solution which freezes at minus 1.86°C and boils at plus .54°C.
9. A molar solution of an electrolyte has a lower freezing point and a higher boiling point than has the same concentration of a solution of a non-electrolyte.

C. *The Theory of Ionization*

1. First advanced by a Swedish chemist, Arrhenius, in order to explain certain facts which hah been observed.

   (a) Why does a given solution of an electrolyte have a lower freezing point, a higher boiling point, and a higher osmotic pressure than does a solution of a non-electrolyte of the same concentration?

(b) How does a solution of an electrolyte conduct an electric current?

(c) What is the explanation of such chemical reactions as these:

1. The hydrolysis of aluminium sulphate in solution gives a solution having an acid reaction, while the same reaction involving sodium carbonate gives a solution having a basic reaction?
2. Why do certain reactions, such as neutralizatian of an acid by a base, go to completion?

2. The ions are formed in a compound when the metallic radicals give up their valence electrons to non-metallic radicals, forming the cations and the anions. Actual dissociation generally does not take place except in solution in water, or, in certain cases, other solvents.

D. Colloids

1. Colloids consist of a dispersed phase in a dispersion medium. Illustrations of colloids are clouds, fog, smoke, dust in the air, milk, gelatin in solution, emulsions, coloured glass, etc.
2. One can form a colloidal solution of ferric hydroxide by making a saturated solution of ferric chloride in distilled water, and then pouring this solution into boiling distilled water. Study its properties. Why is this solution stable?
3. A good example of an emulsion can be made by placing a small amount of kerosene (or a light oil) water and then shaking, after adding several milliliters of soap solution or liquid soap. The soap is an emulsifying agent.

4 In such an industry as a cement mill, the colloidal dust in the air creates a major problem that can be solved by means of the Cottrell precipitator. How does it work?

5. The blood, lymph, and many body tissues are colloidal in nature.

IV. Learning Activities

A. Teaching Suggestions

1. Review the electron theory covered in Unit II.
2. Bring out similarities and differences between atoms and ions, and show how ions are formed.
3. Arrange an apparatus for electrolysis, using an electric light bulb to indicate the presence of a current passing through the system, and measure the conductivity of both dilute and concentrated solutions of several substances such as sodium chloride, the common acids and bases, cane sugar, ethyl alcohol, mercurochrome in distilled water, etc.
4. Develop the problem of man's efforts to supply himself with a safe, adequate water supply. Cover all of the natural sources.
5. It will be interesting to point out that the largest spring in the United States in Silver Spring, near Ocala, Florida, and the second largest is Big Spring, near Van Buren, Mo., which has an estimated daily flow of 300,000,000 gallons.
6. Make a study of the physical and chemical properties of water. Much of this can be developed directly from the laboratory work.
7. Make a study of a municipal water purification plant. After studying the principles involved, visit such a plant if possible.

8. Make an experimental study of hardness in water, causes, and methods of removal of hardness. How is hardness related to the amount of soap used? How is it related to the use of boiler feed water?
9. Bring out what happens when a solute dissolves in a solvent. Illustrate by the use of both a liquid and a solid as a solute.
10. Develop the concepts of saturated and super-saturated solutions, basing the discussion on experiments.
11. Have the pupils write out specific directions for preparing molar, molal, and normal solutions of several common substances, such as sodium chloride, hydrated cupric sulphate, and sulphuric acid.
12. In developing the concept of the theory of ionization, show the reasons for the theory, and show how the theory explains the abnormal freezing point, boiling point, and osmotic, pressure of electrolytes, as well as the mechanism of conductivity of the electrolytic current through and electrolyte. Show also how this theory explains the cause of chemical changes.
13. Develop the concept of colloids, bringing out the fact that the colloid may be a gas, a liquid, or a solid, and that, in each case, there is a dispersed phase and a medium in which this colloidal material is dispersed.

    (a) Colloidal particles carry electric charges which which come from adsorbed ions. This electric charge is the reason why many colloids are stable.

    (b) In such colloids as oil-in-water emulsion, soap is an emulsifying agent. Bring out the fact that eggs have the same function in mayonnaise and salad dressings.

(c) Such colloids as gelatin solutions are stable partly because of the charge of the particle, but also because the particles are highly hydrated.

(d) The size of colloidal particles is generally between 5 and 100 millimicra. (A millimicron is a millionth of a millimeter.) If particles are larger they tend to settle out, while those that are smaller are approaching molecular magnitude.

(e) In developing those concepts, bring out the important industrial applications of colloids, such as the relation of the colloidal state to dust explosions, the operation of the Cottrell precipitator, the smoke nuisance, fog, clouds, emulsions, etc.

B. Pupil Activities

1. The following series of tests for potability (suitability for drinking purposes) of water may be done by pupils under careful supervision. It is well to have the pupil furnish his own water sample from a known source, and, after the tests have been completed, be prepared to present his findings, together with a description of the well, for example, from which the water was obtained. If several samples of water are run through this series of presumptive tests, one can acquire enough experience to be able to draw some conclusions as to the contamination, if any. It should be remembered, in this connection, that in the breaking down of protein in organic matter, the products formed in order are ammonia, nitrites, and nitrates, and from that may be inferred the length of time during which oxidation has taken place. Thus, a sample that contained only ammonia has probably

a more immediate source of organic matter than has one containing only nitrates. The tests that are suggested follow:

(a) Tests for the potability of water

In the analysis of water no one test may be considered conclusive. The chemical tests listed herein are indicative of a condition in the water that may form the basis for conclusion as to the probability of the water's being contaminated.

In such analysis five tests are often made. These tests are for the following named substances:

(a) *chlorides*,
(b) *organic matter*,
(c) *nitrates*,
(d) *nitrites*, and
(e) *ammonia*.

(b) Tests for chlorides

Take a test tube of water to be tested and add a few drops of a dilute solution of silver nitrate. Acidify with a few drops of nitric acid. A white precipitate or distinct opalescence indicates the presence of chlorides in the water. The degree of precipitation is a measure of the chlorides present. Chlorides in water may mean sewage contamination, or chlorides from salt beds in the ground, or other mineral source.

(c) Test for organic matter

Take a test tube of water to be tested and add 2 or 3 ml., of concentrated sulphuric acid. Then add a few drops of a very dilute solution of potassium permanganate so as to produce a very light violet colour. Devide the sample into

2 equal parts and boil one portion gently for a few minutes. If organic matter is present in the water the colour of the permanganate will fade. The degree of fading indicates the amount of organic matter. If solution turns brown the test may be repeated using more acid. The heated portion should be compared with the portion that was not heated in order to judge the amount of fading of colour. Organic matter may be of sewage origin or it may result from decaying animal or vegetable matter.

(d) Test for nitrates

Take a test tube of water to be tested and add about 0.1 of a gram of ferrous sulphate, or of ferrous ammonium sulphate. Then, holding the tube in an inclined position, pour cautiously down the side of the tube about 5 ml., of concentrated sulphuric acid in such manner that the two liquids do not mix. If nitrates are present a brown ring will form at line of contact of the two liquids. In making this observation the observer should turn his back to the widow and hold a white card or paper in such manner as to reflect the light from the card through the liquid. This will exclude many chances for error.

(e) Test for nitrites

Take two test tubes of water to be tested and to one add a portion of Meta phenylene diamine hydrochloride about the size of a common pinhead. Let stand for five minutes. If nitrites are present a yellow colour will develop. The intensity of the colour produced may be judged by comparison with the untreated sample. In making this observation both tubes should be grasped in the hand in such a way as to exclude the extraneous light, taking care to place the treated portion where it can be easily recognized, then taking a position so that light from a card or paper may be reflected up three the length of the tubes.

(f) Test for ammonia

To a test tube of water to be tested add about 1 ml., of Nessler's solution by means of a pipet, taking care not to stir up any residue in the bottle of the reagent. In case ammonia is present a yellowish-brown colour will develop. This should be examined as in the case of the test for nitrites, using light reflected from a white card.

C. *Other Experiments*

1. An experiment should be planned that will show the fact that some crystaline substances are hydrates and others are not.
2. As an optional experiment, the pupils could design an experiment in which the percentage of water of hydration in a hydrate is determined.
3. An experiment should be planned that will show the phenomena of deliquescence and efflorescence, using, for example, calcium chloride and hydrated sodium sulphate.
4. A series of experiments may be planned to show the total hardness in water, testing such compounds as sodium carbonate, borax, and soap as to their ability to remove the hardness.
5. An interesting pupil demonstration experiment is the electro lysis ot water.
6. An interesting experiment designed to show the effect of certain conditions upon the rate of

solution of a solute may be conducted as follows: Take two samples of cupric sulphate of the same mass, one powdered and the other in large crystals, and place each in the bottom of a test tube filled with distilled water. At the same time place two similarly prepared samples of the same salt suspended in filter paper just beneath the surface of water in each of two other test tubes. Watch the rates ol solution, and note the order in which the solute completely goes into solution. What does this show as to the effect of these two factors on the rate of solution?

7. Prepare a saturated solution of such a compound as sodium sulphate and study its properties, as is outlined in any good laboratory manual. This might be a good group demonstration experiment.

8. As a group or individual pupil demonstration experiment, prepare a super-saturated solution of such a compound as sodium nitrate. To do this best prepare a saturated solution at about 70 to 80 degrees, filter hot, and let the filtrate cool without agitation or stirring. When the temperature has dropped to room temperature drop, a crystal of the original solute into the solution and watch the crystallization take place.

9. As an optional experiment, the teacher might want to prepare a normal solution of sodium hydroxide and a similar solution of hydrochloric or sulphuric acid, and ascerta in the relative volumes of each of the two solutions required to produce neutrality, using some such indicator as litmus solution or methyl orange. The determination of an unknown acid or base could be done after the solutions have been standardized.

10. An interesting experiment showing ionic migration can be devised using cupric sulphate in a gelatin solution in a U-tube. After the gelatin has

hardened, place a gelatin sulution of barium chloride in one side of the U-tube above the cupric sulphate solution, and an ammoniacal solution of gelatin in similar position in the other side of the tube. Connect the positive pole of a battery to a small metal plate and insert the plate in the end of the tube containing the ammonium hydroxide. Connect a similar metal plate to the negative pole of the battery, and insert the plate in the other end of the U-tube. When the current has been flowing for some time the tendency of the cupric ions to migrate toward the negative pole can be seen in the dark blue colour appearing near that pole while the migration of the sulphate ion can be seen by the appearance of barium sulphate precipitate appearing in the other end of the tube.

11. A good pupil demonstration experiment can be arranged using any source of direct current to show the relationship between degree of ionization and the ability of a solution to conduct an electric current. If a 100-volt current is used, connect an electric lamp in as resistance. It might be well to have a committee of the class plan this experiment and have their plan approved by the instructor before they set it up for a class demonstration.

12. The pupils individually, or as a group, should prepare a colloidal solution of ferric hydroxide by taking a saturated solution of ferric chloride in distilled water, and pouring it into about 50 ml., of boiling distilled water. To show the flocculation of this colloid, prepare a molar solution of sodium chloride, one of sodium sulphate, and one of sodium phosphate, and add each of these solutions separately, drop-wise, to separate 10 ml., portions of the colloid. Study the relative amounts of the three electrolytes needed to produce beginning of

flocculation in each tube. Which ion has the greatest tendency to produce flocculation, the chloride ion, the sulphate ion, or the phosphate ion?

13. Prepare an oil-in-water emulsion by mixing about 0.5 ml., of kerosene or a light oil with about 10 ml., of water using soap solution as an emulsifying agent. Does mayonnaise constitute a similar type of colloid?
14. The pupils or the teacher might well prepare a demonstration on the Tyndall effect. To do this focus a beam of light through a colloidal solution, and note the path of the beam through the solution. A car driving through a foggy night is an equally good demonstratian of this phenomenon.

D. Optional Work

1. Experiments dealing with diffusion through a collodion bag. The colloidal solution of ferric hydroxide may be purified by this means. The process is known as dialysis.
2. The preparation of a calcium acetate gel, and the study of its properties.
3. The making of gelatin gels; the use of pectin in making fruit jellies.
4. The preparation of any one of several types of colioidal systems. In this the pupil should work out his plan, and secure the approval of the teacher in advance of beginning the experiment.
5. The preparation of quick-setting Plaster of Paris.
6. Demonstration or lecture on the Cottrell precipitator.
7. A report on the flotation process of separating certain ores from the rock mass with which they occur.

8. A 500-word theme on any one of the following named topics:

    H-ion Concentration in Industry

    The Elementary Principles of H-ion Control

    Glauber's Salt

    The Manufacture and Uses of Sulphuric Acid, or of Caustic Soda.

    Salts in Industry

    Mining of Salts

NOTES

A. 500-Word Themes

1. H-Ion Concentration in Industry
   "The A.B.C. of H-ion Control," La Motte Chemical Products Co., Baltimore, Md.
2. Glauber's Salt
   Scicnce News Letter, June 7, 1930.
3. Uses of Caustic Soda
   Encyclopedia, *Chemistry Leaflet*, October 15, 1931.
4. Salts in Industry
   Chemistry Leaflet, October 15, 1931.
5. Mining of Salt
   Ramance of Modern Mining by A., Williams, Chap. 13, Seely Service Co., Ltd., 38 Great Russell St., London, England.

B. *Text Materials*

1. Biddle and Bush, *Dynamic Chemistry*, Rand McNally, 1937, pp. 161-206; 313-578.
2. Black and Conant, *New Practical Chemistry*, Macmillan, 1936, pp. 372-393.
3. Brauer, *Chemistry and Its Wonders*, American, 1938, pp. 622-634.
4. Brownlee et al., *Chemistry in Use*, Allyn and Bacon, 1939, pp. 116-137.

5. Brownlee et al., *First Principles of Chemistry*, Allyn and Bacon, 1940, pp. 45-64; 452-457.

6. Bruce, *High School Chemistry*, World Book Company, 1938, pp. 172-223; 448-454.

7. Dull, *Modern Chemistry*, Henry Holt, 1940, pp. 114-129; 344-362; 473-490.

8. Fletcher, Smith and Harrow, *Beginning Chemistry*, American, 1929, pp. 13-156; 78-104.

9. Gray, Sandifur, and Hannah, *Fundamentals of Chemistry*, Houghton Mifflin, 1938, pp. 87-95; 582-594.

10. Greer and Bennett, *Chemistry*, Allyn and Bacon, 1939, pp. 38-62.

11. Hopkins et al., *Chemistry and You*, Lyons and Carnahan, 1939, pp, 113-138; 401-424.

12. Horton, *Modern Everyday Chemistry*, Health, 1937, pp. 219-234.

13. Howard, *Units in Chemistry*, Henry Holt, 1938, pp. 71-122.

14. Jaffe, *New World of Chemistry*, Silver Burdett, 1940, pp. 235-250; 638-649.

15. Kruh, Carleton, and Carpenter. *Modern-Life Chemistry*, Lippincott, 1937, pp. 135-182.

16. McPherson, Henderson, and Fowler, *Chemistry at Work*, Ginn. 1938, pp. 63-78; 410-427.

CHEMISTRY

TEACHER EVALUATION AND SUGGESTIONS FOR IMPROVEMENT

Unit ....
(*Number*) (*Title*)

Note to teacher: As a result of your experiences in teaching this unit, please make suggestions for improvement and send to Lloyd W. King, State Superintendent of Public Schools. Use an additional sheet if necessary.

Specific Objectives

Sucject Content

Learning Activities

12

Unit V
The Metals and Their Compounds

I. Specific Objectives

A. To acquaint pupils with the chemical and physical properties of metals and their compounds.

B. To acquaint pupils with the uses of metals and their compounds.

II. Vocabulary

metamorphic	slaking	spiegeleisen
sedimentary	calcining	open-hearth furnace
igneous	gangue	crucible
magna	matte	alloy
malleability	slag	tensile strength
ductility	flux	welding
conductivity	charge	annealing
ceramic	flotation	corrosion
plasticity	dressing	galvanizing
minerals	smelting	tinning
rocks	roasting	flame test

mineralogy	blast furnace	bead test
Solvay process	fusion	Bessemer converter
metallurgy	reverberatory furnace	

III. Subject Content

A. *General—Physical and Chemical Properties of the Metals; Mineralogy and Metallurgy*

1. Metals may be contrasted with non-metals on the basis of their physical properties of luster, malleability, ductility, conductivity, and tensile strength.
2. The chemical behaviour of metals varies markedly, all "lend" electrons as contrasted with the non-metals, which "borrow" electrons.
3. Most familiar metals occur in combination. Their ores are usually oxides, sulphides, or carbonates. Gold, platinum, and a few otner rare metals are usually found in the free state.
4. The extraction of metals from their ores is called metallurgy, Metallurgical methods include:

   (a) Simple heating, as in the extraction of mercury.
   (b) Roasting or calcining the sulphide or carbonate to an oxide.
   (c) Reduction of the oxide by carbon or aluminium.
   (d) Electrolysis.
   (e) Replacement in solution.

B. *Iron and Steel, Nickel, and Cobalt*

1. The importance of iron to modern civilization can hardly be overestimated as it touehes practically every phase of our lives.

2. Cast or pig iron is produced in the blast furnace.
3. Wrought iron is produced in the reverberatory furnace. It is the purest form of iron, and can be forged and welded.
4. Low grade steel is produced by the Bessemer converter. High grade steel is obtained from the open-hearth furnace, by the crucible process, and from the electric furnace.
5. The tempering of steel depends in part upon the changing of iron from one allotropic form to another.
6. Useful steel alloys are obtained by adding other metals such as nickel, chromium, manganese, vanadium, molybdenum, etc., in the proper proportions.
7. Iron forms two series of compounds: ferric, in which it is tervalent, and ferrous, in which it is bivalent. The tests for these ions involve the formation of the ferrocyanides and ferricyanides.
8. The ferrous ion may be oxidized to the ferric, while the ferric ion may be reduced to the ferrous. These reactions may or may not involve oxygen.

   (a) Blueprinting involves the reduction of ferric ions in the presence of light and a suitable reducing agent, and the subsequent formation of ferrous ferricyanide.

   (b) The change in colour observed in blue-black ink is due to the oxidation of ferrous ions and the precipitation of black ferric tannate and ferric gallate.

9. Cobalt finds some use in colouring glass blue and in making magnets for telephone coils. Cobalt nitrate is used in tests for aluminum and zinc.
10. Nickel is used in plating iron, in making stainless

steel, in minting of "nickels," in nickel-chromium wire in electric healting devices, and in Monel metal.

C. *Aluminium*

1. Charles Martin Hall, at the age of twenty-two, discovered an inexpensive process of extracting aluminium, thereby creating a great industry, reducing the cost of aluminium from four dollars to twenty cents a pound, and gaining a fortune of forty million dollars.
2. The extraction of aluminium illustrates metallurgy by electrolysis.
3. The uses of aluminium depends largely on its lightness, conductivity, and resistance to tarnish.
4. Aluminium hydroxide has interesting properties and uses.
   (a) It is amphoteric.
   (b) The dyeing industry makes use of it as a mordent.
   (c) It is used as a coagulant in water purification.
5. Aluminium is a powerful reducing agent. The thermit process depends upon this property of aluminium.
6. Alums are double salts.
7. The development of a cheap method of producing aluminium chloride for use in cracking oils illustrates the methods of modern industrial research.

D. *The Alakali Metals*

1. The compounds of sodium and potassium occur abundantly in the earth's surface.

2. There are numerous processes for obtaining various useful compounds of the metals. One of the most important of these is the Solvay process.
3. The chemistry of sodium and potassium is not striking, but the economic importance of their compounds is very great.
4. Rubidium and caesium find some use in photo-electric cells.

E. *Calcium, Strontium, and Magnesium*

1. The calcium compounds are very important in the industries and arts.
   (a) Limestone is used in the production of quicklime, slacked lime, mortar, cement, and concerete.
   (b) Calcium sulphate is used in the production of Plaster of Paris and treating "black alkali" land.
   (c) Calcium chloride, is an important drying agent.
   (d) Chloride of lime is a useful bleaching agent and disinfectant.
   (e) The calcium phosphates are used as fertilizer, in baking powders, and in medicine.
2. The calcium and magnesium compounds in hard waters precipitate soap, thus wasting it and producing little gray spots in clothes that are laundered in it. They also form "boiler scale," resulting in economic loss. Waters containing the bicarbonates of these metals can be softened by boiling. Waters containing other salts of these metals may be softened by treating chemically.
3. Strontium salts are used to some extent in red fired and in fireworks.
4. Barium compounds are used in X-ray photography of the alimentary canal.

F. *Other Metals*

The uses, occurrence, and properties of copper, zinc, mercury, tin, silver, gold, chromium, and lead are suggested under *Learning, Activities.* These should be handled as time and available material permit, with particular emphasis on the uses.

IV. Learning Activities

A. *Teaching Suggestions—Emphasize Development of Problem Consciousness.*

1. Recall and review questions, examples of which are:

    (a) What natural rocks are used as building materials?
    (b) What is the origin and method of formation of limestone?
    (c) List as many uses for limestone as you can.
    (d) What do you know about the formation of minerals in the earth?
    (e) What are some uses of common salt?
    (f) Give an electronic explanation of oxidation and reduction; of single replacement.
    (g) What are some ways in which metals are protected against corrosion?
    (h) Prepare a list of uses of metals in the home; in building; in transportation.
    (i) Why is our present civilization sometimes called the Age of Steel?
    (j) Might it be that we are entering the Age of Aluminium? Why?

2. Exhibition of available minerals, metals, and alloys
3. Report: "The Dependence of Modern Civilization on Metals"

4. Demonstrations designed to arouse student interest and curiosity, such as:

   (a) Blueprinting
   (b) "Invisible inks"
   (c) Mosaic gold
   (d) Reduction of a metallic oxide by aluminium

B. Pupil or Teacher Demonstrations: Accompanied by Verbal Explanations

1. General

   (a) Roasting and reducing cupric sulphide by carbon
   (b) Obtaining copper by electrolysis
   (c) Thermit process

2. Aluminium

   (a) Formation and properties of aluminium hydroxide
   (b) Aluminium hydroxide as a coagulant in the purification of water
   (c) Preparation of alums

3. Calcium, strontium, and magnesium

   (a) Limestone, quicklime, and slaked lime, including the carbon dioxide test
   (b) Mortar and cement
   (c) The different kinds of hard water
   (d) Action of soap on hard water
   (e) Softening hard water
   (f) Strontium fire (Use potassium chlorate and strontium nitrate bound with orange shellac.)

4. Other metals

   (a) Preparation of cuprous and cupric compounds
   (b) Tests for cuprous, cupric, mercurous, mercuric, zinc, tin, lead, silver, and chromium ions

(c) Zinc amalgam
(d) Preparation of mercurous and mercuric salts
(e) Galvanizing iron
(f) Preparation of various pigments
(g) Effect of light on silver salts
(h) Silvering a mirror
(i) Chromates and dichromates

C. Experiments

1. General

(a) Reduction of lead monoxide by carbon with blow-pipe or glass tube
(b) Roasting of lead sulphide and reduction of oxide
(c) Recovering copper by replacement

2. Iron and steel, nickel and cobalt

(a) Tests for ferric and ferrous ions
(b) Oxidation of ferrous and reduction of ferric ions with identifying tests
(c) Iron inks—formation of ink and removal of stains

3. The alkali metals
(a) The Solvay process
(b) Baking powders

D. Additional Activities

1. General reports

(a) Descriptions of mines, metallurgical processes, manufacturing processes, or research involving metals that may have been observed by any of the students
(b) The invention and application of the spectroscope
(c) Crystallograpay

2. Iron and steel, nickel and cobalt (reports)

(a) Useful steel alloys
(b) The structure of steel (including the physical and

chemical changes involved in the hardening and tempering of steel.)

(c) Crucible steel and the electric refining of steel

(d) Nickel and cobalt

3. Aluminium (reports)

   (a) The life of Hall and his contributions to chemistry
   (b) Gems
   (c) Aluminium alloys and their uses in various industries
   (d) Arguments for and against alum baking powders
   (e) The manufacture and care of kitchen utensils
   (f) Ceramics
   (g) The history of the development of aluminium chloride in cracking oils

4. The alkali metals (reports)

   (a) The story of salt
   (b) The less common alkali metals

5. Calcium, strontium, and magnesium (reports)

   (a) The story of cement
   (b) Stalactites and stalagmites

6. The less common metals (reports)

   (a) The prevention of corrosion

      (1) Tinning and zincing
      (2) Paints and pigments

   (b) Photography
   (c) Minting
   (d) Electroplating
   (e)
   (f) The life of Madame Curie
   (g) The development of the electric light bulb filament

7. Field trips

   Trips to any available mines, metallurgical or manufacturing plants, or research laboratories that would help to develop an understanding of the material included in this unit would be of value.

8. Socialized procedure and group experimentation

   (a) Suggested laboratory experiments that can best be worked out by dividing the group into committees include:

      (1) The general physical properties of the metals. Have each committee determine qualitatively as far as is practicable the luster, conductivity, malleability, hardness and density of one of such commonly available metals as copper, zinc, lead, iron, tin, aluminium, and mercury.
      (2) The electrochemical series. Have each committee work with only one metal to determine which metals it will or will not displace from its salts.
      (3) The action of metals towards acids and bases.

   (b) If desired, the entire unit could be handled on a committee plan by assigning to each committee one of the topics indicated by capital letters under *Subject Content*.

9. Projects. The projects that are starred in the following list should probably be carried out by each student. The others may be entered upon as time permits and pupil interests directs.

   (a) General

      (1) Make a mineral collection on your field or pleasure trips.
      (2) By using references, outline a procedure and analyze as far as possible various commercial products, such as cosmetics, soaps, dentrifices, parent medicine, etc. The detection of possible harmful metals or acid redicals could be made the basis of a report.

   (b) Iron and steel, nickel, and cobalt

(1) Make labeled diagrams of the blast furnace, the open-hearth furnace, and the Bessemer converter. It is suggested that this be accompanied by papers giving details of operation and the reactions involved.

(2) Make blueprints of flat objects such as keys, leaves, etc., or from drawings on transparent paper.

(3) Construct a small working model of a blast furnace.

(4) Make a cobalt chloride hygrometer of a novel design.

(c) Aluminium

(1) Construct a working model to show thermit welding.

(2) Try the effect of various mordants and dyes on various materials and mount your samples on a poster.

(3) Make a labeled diagram of the Hall apparatus and explain the chemistry involved.

(d) The alkali metals

*(1) Prepare a chart or table showing the formulas, important mineral sources or method of preparation, and important uses of the common compounds of sodium and potassium.

*(2) Prepare a flow chart showing the different derivatives that may be prepared from sodium chloride.

*(3) Prepare a flow chart showing the details of Solvay Process.

(4) Construct a model soda-acid fire extinguisher.

(e) Calcium, strontium, and magnesium

*(1) Prepare a flow chart showing calcium carbonate as the source of other calcium compounds.

*(2) Prepare a table or chart showing the chemical names, common names, formulas, and important uses of calcium and magnesium compounds.

(3) Starting with a lump of limestone, prepare as many different compounds as possible.

(4) Analyze samples of "boiler scale."

(5) Construct a small working model of a water softening plant.

(6) Construct a small model of a cement plant.

(7) Make Plaster of Paris casts of various objects.

(8) Make a series of "soap suds tests" to determine the effectiveness of various commercial water softeners and report on these to the class.

(f) The less common metals

(1) Try etching designs on copper.

(2) Make some Pharaoh's serpent eggs from mercuric thiocyanate. Danger from fumes!)

NOTES

A. General

Foster, William, *The Romance of Chemistry*, Appleton-Century, 1936, pp. 261-280.

Furnas, C.C., *The Storehouse of Civilization*, Columbia University, 1939.

Clarke, Beverly L., *Marvels of Modern Chemistry*, Harpers, 1935, pp. 239-251.

Howe; H.E., *Chemistry in Industry*, Vol. I., Chemical Foundation, 1924-25.

Howe, H.E., *Chemistry in Industry*, Vol. II., Chemical Foundation, 1924-25.

Weeks, Mary E., *The Discovery of the Elements, Revised*, Journal of Chemical Education, 1939.

Howard, "Emeralds," *Journal of Chemical Education*, Vol. II, (June, 1934).

Sadtler, S.S., *Chemistry in Familiar Things*, Lippincott, 1937, pp. 133-140.

B. Texts

1. Biddle and Bush, *Dynamic Chemistry*, Rand McNally, 1937, pp. 355-368; 433-455; 369-388; 398-404; 456-498.
2. Black and Conant, *New Practical Chemistry*, Macmillian, 1936, pp. 476-489; 490-515; 339-355; 394-420; 516-540.
3. Brauer, *Chemistry and Its Wonders*, American, 1938 pp, 239-255; 289-309; 328-337; 256-270; 271-289; 310-347.
4. Brownlee et al., *First Principles of Chemistry*, Allyn and Bacon, 1940, pp. 506-541; 567-611; 542-566; 611-660.
5. Bruce, *High School Chemistry*, World Book Company, 1938. pp. 375-382; 421-438; 454-467; 383-396; 397-413; 476-491; 413-420.
6. Dull, *Modern Chemistry*, Henry Holt, 1940, pp. 491-508; 591-623; 562-576; 509-527; 528-561; 624-666.
7. Gray, Sandifur, and Hannah, *Fundamentals of Chemistry*, Houghton Miffln, 1938, pp. 22-35; 408-418; 389-406; 383-385; 316-329; 385-388; 330-336.
8. Greer end Bennett, *Chemistry*, Allyn and Bakon, 1939, pp. 743-775.
9. Hopkins et al. *Chemistry and You*, Lyons and Carnahan, 1939, pp. 645-653; 685-716 ; 654-670; 679-685; 671-679; 717-729.
10. Howard, *Units in Chemistry*, Henry Holt 1938, pp. 431-446; 515-550; 494-503; 446-453; 479-492; 503-515; 550-612.
11. Jaffe, *New World of Chemistry*, Silver Burdett, 1940, pp. 392-410; 427-450; 411-426; 464-488.
12. Kruh et al., *Modern-Life Chemistry*, Lippincott, 1937, pp, 437-446; 488-508, 447-466; 467-487.
13. Wilson, *Descriptive Chemistry*, Henry Holt, 1936, pp. 281-300.
14. Silson and Mullins, *Applied Chemistry*, Henry Holt, 1939, pp. 471-517.
15. McPherson et al., *Chemistry at Work*, Ginn, 1938, pp 499-517; 572-588; 549-562; 518-527; 528-548; 589-636.

CHEMISTRY

TEACHER EVALUATION AND SUGGESTIONS FOR IMPROVEMENT

Unit..

(Number) (Title)

Note to Teacher: As a result of your experiences in teaching this unit, please make suggestions for improvement and send to Lloyd W. King, State Superintendent of Public Schools. Use an additional sheet if necessary.

Specific Objectives

Subject Content

Learning Activities

13

Unit VI
Useful Non-Metals

I. Specific Objectives

A. To assist pupils to acquire an understanding of the properties, reactions, and electron structure of the important non-metals: sulphur, nitrogen, phosphorus, the halogens, silicon, and boron.

B. To enable pupils to gain knowledge of the uses of each of the nonmetals and its compounds in relation to health, indestry, agriculture, the home, medicine, and national defense.

II. Vocabulary

gypsum	aqua regia	guano
Frasch process	ammonia	white phosphorous
allotropic	nitrate	red phosphorous
rhombic	quartz	phosphate
monoclinic	silicia gel	slag
amorphous	carborundum	halogen
flowers of sulphur	nitrite	bromine water
milk of sulphur	nitrogen fixation	tincture of iodine

roll sulphur	nitrifying bacteria	sublimation
anhydride	refrigeration	silicates
sulphite	legumes	feldspar
sulphate	nitrogen cycle	borax
thiosulphate	saltpeter	borates

III. Subject Content

A. Sulphur and Its Compounds

1. Its occurrence in such volcanic areas as Sicily, and in the United States
2. Methods of mining and refining sulphur
3. The allotropic forms of sulphur
   (a) Two crystalline varieties
      (1) Orthorhombic
      (2) Monoclinic
   (b) Two amorphous varieties
      (1) Plastic
      (2) Milk of sulphur
4. Uses of sulphur
5. The two important oxides of sulphur, sulphur dioxide and sulphur trioxide, and their corresponding acids
   (a) Uses of sulphur dioxide and its properties
   (b) The preparation, properties, and uses of sulphuric acid by both the lead chamber process and the contact proress.
6. Important salts of sulphurous and sulphuric acids, and their uses

B. Nitrogen and Its Compounds

1. The occurrence of elemental nitrogen in nature; the composition of the atmosphere
2. The preparation of pure nitrogen in the laboratory from sodium nitrite and ammonium chloride
   Properties of nitrogen

3. The nitrogen cycle
4. The naturally occurring nitrogen compounds of commercial importance
   (a) Sodium nitrate (Chile saltpeter), and its place in commerce and industry
   (b) Potassium nitrate (saltpater), and its uses
5. The artificial fixation of atmospheric nitrogen
   (a) The Haber process
   (b) The Cyanamide process
   (c) The Are process
6. Preparation, properties, and uses of oxides of nitrogen
   (a) Nitrous oxide, prepared from ammonium nitrate
   (b) Nitric oxide, prepared by the reaction of copper and nitric acid, (one gram copper turnings to 3 ml. dilute nitric acid)
7. Preparation and properties of nitrates and nitrites
8. The preparation, properties, and uses of nitric acid
9. The preparation of ammonia, commercial and laboratory, its uses and properties
10. The use of nitrogen compounds in plant growth and in the maintenance of soil fertility.

C. Phosphorous and Its Important Compounds

1. The natural occurrence of phosphorous compounds and their commercial importance
2. The preparation of the element phosphorus; the properties and uses of red phosphorous and white phosphorous
3. The relation of phosphorous compounds to plant and animal life and to soil fertility.
4. Phosphoric acid, its preparation, and properties

D. The Halogens and Their More Important Compounds

1. Their naturally-occurring compounds
   (a) Chlorine compounds and their commercial importance
   (b) Bromine compounds
   (c) Iodine compounds
   (d) Fluorine compounds
2. The laboratory and commercial preparation of chlorine and its properties and uses
   (a) The uses of elemental chlorine
   (b) The preparation of commercially important compounds of chlorine, such as chloride of lime, phosgene, etc.
3. The preparation of bromine and of iodine, and the properties of each; uses of iodine and bromine compounds
4. The preparation, properties, and uses of hydro-chioric acid
5. The preparation of hydrogen fluoride, and its use in glass etching.

E. Silicon and Boron

1. The important silicon compounds founds in nature
   (a) Silicon dioxide in its various forms
   (b) Feldspars
   (c) Kaolins
2. A brief study of some important silicate rocks and minerals
3. The manufacture of glass
4. The manufacture of various ceramic wares
5. The occurrence, uses, and properties of important boron compounds

IV. Learning Activities

A. Teaching Suggestions

1. Recall in Unit I the reaction of sulphur, phosphorous, and carbon with oxygen. What kinds of compounds was formed when each of the non-metal oxides formed reacted with water? What use did the great French chemist, Lavoisier, make of this observation?
2. In what fundamental respects do non-metals differ from metals? Illustrate this by drawing diagrams showing the atomic structure of such a metal as sodium, and such a non-metal as chlorine or oxygen. List these differences.
3. May certain elements by sometimes metallic and sometimes nonmetallic, depending upon reacting conditions? What term is applied to this property?
4. In the study of each of the non-metals, investigate the source of supply, the form in which it occurs, how it is recovered, and how made ready for use. Is there any problem connetced with any of these essential elements that constitutes an important item in international commerce?
5. In the study of each of the elements in this group, what commercially important compounds are known? If any one is essential to man tell why it is so.

B. Pupil Activities and Demonstrations

1. Teacher or pupil demonstration experiments on sulphur

    (a) Two crystalline and two amorphous forms of sulphur

(b) Preparation of a number of metallic sulphides, and a study of some of the properties of each

(c) Preparation and properties of sulphur dioxide. Prepare the gas by two different methods.

(d) Set up a glass model of a lead chamber sulphuric acid plant, and produce some of the acid. Then demonstrate its properties.

(e) Set up a model of a content process plant for manufacturing sulphuric acid, using a mixture of cupric oxide and ferric oxide precipitated in asbestos fibre as a catalyst. Test the product as to properties. How do the two products of the two plants compare as to purity?

(f) Test for sulphate ion in solutlon, distinguishing it from sulphite ion.

2 In class discussion, develop the physical and chemical properties of sulphur, how it is mined in the United States, and how it is obtained in Sicily. Develop the uses of sulphur.

3. Develop in class the method of preparation, properties, and uses of sulphates, sulphites, and thiosulphates. (Optional)

4. Teachers or pupil demoustration experiments on nitrogen

(a) Devise an experiments to show the percentage of oxygen and of nitrogen in the atmosphere.

(b) Show by heating with powered sodium hydroxide that each of the following named substances contains nitrogen by showing that ammonia is given off: hair, leather, finger nail clippings, pieces of horse's hoof.

(c) Devise a laboratory method of preparing ammonia, and study its properties. From the gas prepare three ammonium salts.

(d) Using $NANO_3$ prepare nitric acid, and study its properties.

(e) Prepare nitrous oxide, nitric oxide, and nitrogen dioxide, and study the properties of each.

(f) Test for the nitrate ion.

5. In class discussion develop our fundamental need of an adequate supply of available nitrogen. Show how this demand was met by the natural deposits in Chile, and then by the three important commercial methods of nitrogen fixation; the Haber process, the Arc process, and the Cyanamide process. What effect is the artificial fixation of nitrogen having on the Chilean industry? Bring out our agricultural uses of nitrogen.

6. Draw a diagram of a nitrogen cycle that will show how elemental nitrogen may be changed into nitrogen compounds, and how they pass through the oxidation cycle from ammonia to nitrates, where they are available for plant use, and then how the decay of plants, or their consumption as food by animals changes the nitrogen, ultimately changing it into ammonia Show denitrification by reduction of nitrates and nitrites,, with consequent loss of available nitrogen. Show what happens when the nitrogenous plant or animal body is burned, so far as nitrogen is concerned.

7. Wherē must animals get their supply of nitrogen as food?

8. Have committees of the class report on the manufacture of such explosives as :

(a) gunpowder;
(b) guncotton;
(c) nitroglycerine;
(d) smokeless powder;
(e) trinitro toluence (TNT). Give as much of the history of each as possible; and bring out the

uses of each. A good encyclopedia will give much information on each of these.

9. Teacher or pupil demonstration experiments on phosphorus

   (a) The burning of white phosphorus in air to produce P_2O_5. Dirsolve this gas to form an acid in water solution. Study these reactions.
   (b) Test for the phosphate ion using one of the common tests
   (c) Prepare phosphine using white phosphorous and a solution of KOH, adding about 5 ml. of ethyl ether to drive out the air from the generator. Show that the phosphine bursts into flame immediately upon the bubbles of the gas bursting on the surface of the water in the pneumatic trough.

10. Develop in class the important commercial deposits of phosphorous compounds, and the uses made in agriculture and industry for such compounds. Are we in this country independent in this regard?
11. Why do animals have to have phosphorous in this food? Where do they get it?
12. Teacher or pupil demonstration experiments on the halogens

    (a) Devise a method for preparing each of the three halogens (chlorine, bromine, and iodine in the pure state, and demoustrate the chemical and physical properties of each.
    (b) Prepare hydrogen chloride, and show its chemical and physical properties. Use NaCl H_2SO_4. Can you prepare pure HBr, or HI by the same method? Give reason for your answer.

(c) Devise a method by which you could prepare pure HBr, or HI, using the the tri halide of phosphorous and water.

(d) Show the reaction of the chloride ion, bromide ion, and iodide ion upon silver nitrate.

(e) Prepare calcium hypochlorite, using calcium oxide and chlorine.

13. Develop the important uses (commercial) of chlorine, of bromine and of iodine, as well as our commercial sources of each.

14. Develop at some length the use of chlorine or chlorine compounds in the purification of water, and the control of the spread of disease.

15. Develop the uses of hydrochloric acid, and its manufacture.

16. Demonstration silicon and boron.

17. Develop the importance of silicon in silicate rocks and minerals. Develop the uses of silicon dioxide in the manufacture of glass.

18. Develop the occurrence and uses of borax, and the use of boric acid.

C. *Pupil Reports*

The following topics are suggested for pupil special reports, or for committee investigation and report:

1. The use of halogens in photography
2. The manufacture of glass
3. The manufacture, history, and uses of cement
4. The manufacture of pottery
5. The manufacture of chinaware
6. The manufacture of explosives
7. Nitrogen fixation

8. The use of halogens in the manufacture of war gases, and methods of combating them
9. Manufacture of matches
10. The history of the discovery of one of the elements
11. The commercial purification of water supplies
12. The biography of Herman Frasch

NOTES

1. Biddle and Bush, *Dynamic Chemistry*, Rand McNally, 1937 pp. 286-305; 234-256; 256-261; 207-224; 306-320.
2. Black and Conant, *New Practical Chemistry*, Macmillian, 1936, pp. 187-216; 262-311; 129-148; 327-338; 410-420.
3. Brauer, *Chemistry and Its Wonders*, American, 1938, pp. 416-432; 133-139; 185-200; 598-612.
4. Bruce, *High School Chemistey*, World Book Company, 1938, pp. 239-262; 263-297; 366-374; 119-130; 340-352; 325-336.
5. Dull, *Modern Chemistry*, Henry Holt, 1940, pp. 375-405; 223-281; 440-455; 406-427; 456-472.
6. Gray, Sandiphur, and Hannah, *Fundamentals of Chemistry*, Houghton Mifflin, 1938 pp. 192-212; 225-249; 249-255; 213-224; 345-356; 419-438.
7. Greer and Bennett, *Chemistry*, Allyn and Bacon, 1939, pp. 708-742; 617-651; 686-707.
8. Howard, *Units in Chemistry*, Henry Holt, 1938, pp. 341-354; 380-403; 403-417; 361-380; 328-340.
9. Hopkins et al. *Chemistry and You*, Lyons and Carnahan, 1939, pp. 263-310; 363-400; 731-732; 311-334.
10. Horton, *Modern Everyday Chemistry*, Health 1937, pp. 412-413; 277-287; 268-287.
11. Jaffe, *New World of Chemistry*, Silver Burdett, 1940, pp. 338-340; 270-294; 518-521; 148-167; 536-557.
12. Kruh, et. al. *Modern Life-Chemistry*, Lippincott, 1937.
13. Wilson and Mullins, *Applied Chemistry*, Henry Holt, 1939.
14. McPherson *et. al. Chemistry at Work*, Ginn, 1938, pp. 348-370; 266-309; 371-376; 218-231; 334-347; 390-410.

CHEMISTRY

TEACHER EVALUATION AND SUGGESTIONS FOR IMPROVEMENT

Unit..
(*Number*) (*Title*)

Note to Teacher: As a result of your experiences in teaching this unit, please make suggestions for improvement and send to Lloyd W. King, State Superintendent of public Schools. Use an additional sheet if necessary.

Specific Objectives

Subject Content

Learning Activities

14

Unit VII
Carbon and Its Compounds

I. Specific Objectives

1. To enable pupils to become familiar with some of the many industrial and individual uses of carbon and its compounds.
2. To show how the element carbon and its compounds must be changed in form and properties before they can be used by man.
3. To cause pupils to appreciate how the many thousand compounds of carbon are specifically named so that there is no confusion in the names.
4. To study how the chemical and physical properties of the forms of carbon and some of its compounds determine their use to man.
5. To assist pupils to acquire a knowledge of the physiological importance of the compounds of carbon.

II. Vocabulary

cracking　　hydrolysis　　protein

isomer	saponification	amino acid
tautomer	emulsion	polypeptide
hydrocarbon	sucrose	proteose
alcohol	glucose	plasticizer
aldehyde	levulose	vitamin
ketone	amylose	hormone
ester	celenese	virus
amine	viscose	pseudo-crystalline
homologue	nylon	

III. Subject Content and Learning Activities

A. Problems and Teaching Suggestions

1. Have the pupils prepare a glossary of unramiliar terms as they are encountered.
2. Development
   (a) Outline of content
   1. Carbon
      (a) Uses
      (b) Occurence
      (c) Preparation
      (d) Propesties
         1. Physical
         2. Chemical.
   2. The compounds of carbon
      (a) With oxygen
         1. Carbon dioxide
         2. Carbon monoxide
         3. Carbonates.
      (b) With hydrogen
         1. Hydrocarbons
         2. Liquid fuels.
      (c) With nitrogen and other elements
         1. Cyanides
         2. Amines
         3. Carbon tetrachloride
         4. Carbon disulphde.

(d) With two or more elements

1. Alcohols, aldehydes, ketones, acids
2. Fats, soaps, cleaning
3. Carbohydrates, starches
4. Cellulose, paper, rayon, plastics
5. Proteins, amino acids
6. Vitamins, hormones, viruses.

(b) Introduction (bringing about motivation)

Inspirational story such as Kekule's day dream of the benzene ring, Moore, *History of Chemistry*, McGraw Hill, 1918, p. 213. There is a distinct interest in Kekule's own account of how the idea first come to his mind:

> "I was busy writing on my textbook but could make no progress—my mind was on other things. I turned my chair to the fire and sank into a doze. Again, the atoms were before my eyes. Little groups kept modestly in the background. My mind's eye, trained by the observation of similar forms, could not distinguish more complex structures of various kinds. Long chains here and there more firmly joined; all winding and turning with a snake-like motion. Suddenly one of the serpents caught its own tail and the ring thus formed whirled exasperatingly before my eyes. I woke as by lightning, and spent the rest of the night working out the logical consequences of the hypothesis. If we learn to dream we shall perhaps discover truth. But let us beware of publishing our dreams until they have been tested by the wakening consciousness."

(c) Questions on unit content

Questions on Part I

1. What are the uses of coal as a fuel, and as a source of carbon?
2. For what is coke better suited than coal?
3. What are the uses of charcoal; activated charcoal?
4. What are the uses of lampblack?
5. What uses do we have for graphite?
6. What uses do men find for the diamond other than for adornment?
7. How is coal found, mined, and prepared for use?
8. Since coke, lampblack, and charcoal occur free in nature, why is it that man must prepare all of these forms of carbon that he uses?
9. Does the naturally occurring graphite meet all of our needs for it or must we prepare most of our supply?
10. How are diamonds recovered and prepared for use?
11. What physical properties do each of these forms of carbon have which make them of value to man?
12. What chemical property of carbod causes it to be so useful?
13. Does the lack of chemical activity of the diamond and graphite increase their usefulness?

Questions for Part II

1. Of what importance are the compounds which are formed when carbon unites with oxygen?
2. How can one explain the infinite number of compounds that are possible with carbon and hydrogen?

3. From what were petroleum and natural gas formed, and how were they preserved for our use today?
4. Look up the physlological properties of the two violent poisons, hydrogen cyanide (prussic acid), HCN, and cyanogen, $(CN)_2$.
5. What are amines?
6. Carbon tetrachloride and chloroform are closely related chemically. Carbon tetrachloride is a good fire extinguisher. Is chloroform?
7. What is the greatest industrial use of carbon disulphide?
8. How could one in the laboratory convert an alcohol first into an aldehyde and then to an acid?
9. Soap is made from soap-grease which is a mixture of fats. How is it then that soap will remove such grease from fabrics or our hands?
10. What function does the carbohydrate of our diet perform? (Viscose rayon).
11. Why do we generally cook meat before we eat it?
12. What relation is there between tobacco mosaic (disease) and the proteins?

(d) Teacher demonstrations

1. Blackboard demonstration of structural formulas.
2. Blackboard demonstration on how hydrocarbons, alcohols, and alkyl halides are named according to the two methods (the Geneva names and the common names).
3. The preparation of diethyl ether. (Most any organic laboratory manual).

4. The preparation of aniline (Most any organic laboratory manual).
5. The preparation of aspirin (as follows):

   In a small flask heat a mixture of 10 g. of salicylic acid, 20 g. of acetic anhydride, and 10 drops of conc sulphuric acid to 50° for 10—15 minutes, using a water bath, Now cool the reaction mixture in an ice bath, with stirring, until it thickens to a crystalline, semisolid mass. Add 150 cc. of water, stir until a thin sludge remains, and filter. The solid product is crude acetylsalicylic acid, or aspirin. Dry the product in the open air, and dissolve in the minimum volume of gently boiling ether, heated under reflux condenser over the steam bath.

   To the resulting solution add an equal volume of light ligroin (petroleum solvent) or petroleum ether, and allow to stand on ice for several hours. Collect the resulting purified crystals on the Buechner filter and dry in the open air. The melting point of pure acetylsalicylic acid is not definite, on account of slight decomposition. Figures around 128° 133° are likely to be obtained.
6. The preparation of one of the phthalein dyes. Fuse together 0.5 gram of each of resorcinol and phthalic anhydride—when cool, dissolve in water—fluorscein.

   The number of teacher demonstrations will depend upon many things, such as the teacher's interest, the equipment at hand, the chemicals at hand, and individual class interests. The demonstrations should be confined to short, more difficult experiments, or those in which the element of danger is greater than that of the high-school level. The above list is given only as a suggestion.

(e) Exhibits

Many industrial companies are willing to supply demonstration materials upon request of the teacher, Refer to the series of educational samples of petroleum products put out by the Standard Oil Company.

B. Pupil Activities and Demonstrations

1. Fractional distillation of crude oil

   Note—A satisfactory still can be constructed from pipe fittings by any plumber. The boiler is constructed from a 12-inch length of 2-inch pipe, cappcd at one end, and with a reducing cell from 2 in. to ½ in. A five-foot length of half-inch pipe serves for a condenser.

2. Low temperature coking of coal

   Note—A few lumps of coal are heated in a test tube equipped with a bent tube and stopper, with a Bunsen burner or an alcohol lamp. Coke is formed as is tar and gas.

3. The preparation of absolute alcohol

   Note—The water may be removed from alcohol by distilling the alcohol from lumps of quick lime.

4. The determination of alcohol in a sample of beer

   Note—500 ml. of beer are neutrnlized with Na_2CO_3 and allowed to stand for an hour. The beer is distilled until 100 ml. of it passes over into the distiliate. A 100 ml. flask is carefully weighed empty, then with 100 ml. of water in it, and lastly with the 100 ml. of distillate.

$$\frac{\text{Weight of distillate}}{\text{Weight of the water}} = \text{approximate specific gravity.}$$

The approximate per cent of alcohol by weight and by volume can be found from the ethyl alcohol specific gravity table in the *Handbook of Chemistry and Physics.* This method may be refined by as much as the instructor cares to do so.

C. *Individual or Group Experiments*

1. Preparation of coke, lampblack, and charcoal

   Note—Coke and charcoal can be prepared by heating coal and wood in test tubes equipped with a stopper and delivery tube. Lampblack can be prepared by burning the gas from ether from a jet well up in a wide mouthed bottle.

2. Comparison of absorbtive power between charcoal and activated charcoal.

   Note—Adilute solution of methylene-blue can be used with the two kinds of charcoal for the comparison.

3. Proximate analysis of coal.

   Horton, *Laboratory Manual in Chemistry*, pp. 34-35.

4. Preparation of methane, ethylene, and acetylene

   Davis et al., *Chemistry and You in the Laboratory*, pp. 149-52. Powers and Johnson, *Workbook in Chemistry*, pp. 295-298. Ames and Jaffe, *Laboratory and Workbook Exercises in Chemistry*, pp. 163-166.

   McGill and Bradbury, *Chemistry Guide*, pp. 265-266.

5. Physical properties of alcohols

   Davis et al., *Chemistry and You in the Laboratory*, pp. 153-156.

McGill and Bradbury, *Chemistry Guide*, pp. 269-272.

6. Preparation of ethers

Davis et al., *Chemistry and You in the Laboratory*, pp. 156-158.

Brownlee et al., *Laboratory Experiments in Chemistry*, pp. 245-246.

McGill and Bradbury, *Chemistry Guide*. pp. 273-274,

7. Tests for carbohydrates

Davis et al., *Chemistry and Yon in the Laboratory*, pp. 161-162.

Brownlee et al., *Laboratory Experiments in Chemistry* pp. 253-254.

Powers and Johnson, *Workbook in Chemistry*, pp. 305-306.

McGill and Bradbury, *Chemistry Guide*, pp. 281-284.

D. Supplementary Student Reports

1. Oral presentation of written reports

(a) The story of the diamond
(b) The story of oil
(c) The story of coal
(d) The story of rayon
(e) The story of plastics
(f) What we should know about vitamins
T. G. Archibald, *Journal of Chemical Education*. Vol. 11, 1934. pp. 601-608.
(g) Some interesting things about viruses
W. M. Stanley, *Science*, Vol. 81, 1935, p. 644.

2. Oral reports

(a) My visit to an oil refinery

(b) My visit to a coal mine
(c) My visit to a gas plant
(d) My visit to a research laboratory.

E. *Additional Experiments and Activities*

1. Committee I

(a) The preparation of a soda-acid fire extinguisher
Horton, *Laboratory Manual in Chemistry*, pp. 23-24,

(b) The preparation of nitro cotton and collodion
Horton, *Laboratory Manual in Chemistry*, pp. 52-53.

(c) The preparation of acetaldehyde
Davis et al., *Chemistry and You in the Laboratory*, pp. 153-156.

2. Committee II

(a) The determination of the sweetness of glucose

(b) The preparation of viscose rayon
Place a small piece of cotton in a 100 ml. breaker and cover it with a 20 per cent NaOH solution. Warm the well-stirred cotton for ten minutes. Remove the cotton and squeeze out as much of the solution as possible (wash fingers at once). Place the treated cotton in a test tube, cover it with carbon bisulphide, stopper, and allow to stand for at least 24 hours. It will become orange in colour. Again squeeze the treated cotton as dry as possible; permit the carbon bisulphide in the cotton to evaporate, and then introduce it into a 10 per cent solution of NaOH solution. Stir and shake until it becomes a homogenous viscous mixture. Blow some of the mixture from a tube which has one end pulled down to a fine capillary into a 5 per cent solution of HCl. The thread will be large and irregular but when dry will look like rayon.

(c) The preparation of acetone from calcium acetate,

Introduce 25 grams of calcium acetate into a large thick walled pyrex tube. The tube is fitted with a one-hole stopper and connected to a water cooled condenser by means of a bent glass tube. The test tube is clamped in a horizontal position with the clamp covering the stopper. Apply a flame directly to the test tube beginning near the portion covered by the clamp and gradually working back toward the other end of the tube. Continue heating as long as the liquid distills over. The product prepared is quite impure.

3. Committee III

(a) The preparation of soap

Davis et al., *Chemistry and You in the Laboratory*, pp. 159-160.

Brownlee et al., *Laboratory Experiments in Chemistry*.

Powers and Johnson, *Workbook in Chemistry*, pp. 303-304.

Ames and Jaffe, *Laboratory and Workbook Units in Chemistry*, pp. 169-172.

(b) The testing of unknown samples of cloth

Horton, *Laboratory Manual in Chemistry*. pp. 72-79.

MeGill and Bradbury, *Chemistry Guide*, pp. 289-290.

(c) Preparation of Bemberg silk

Cellulose may be dissolved in Schweitzer's Reagent and reprecipitated by adding the solution to dilute HCl. Schweitzer's Reagent is prepared as follows: Add 5 ml. of a 10 per cent solution of NH_4Cl to 100 ml. of a 5 per cent solution of $CuSO_4$, then add a 10 per cent solution of NaOH as long as the precipitate forms. Wash the precipitate by decantation with several large volumes of water 500 ml.). Filter through cotton cloth and wash until free from the sulphates. Dissolve this freshly prepared $Cu(OH)_2$ in 10 ml, of NH_4OH solution (Conc.). Dissolve sheets of filter paper in this reagent. Blow a very fine stream of the solution from a tube which has been drawn out to a fine tube into a 5 per cent solution of HCl. The commercial name for this fibre is Bemberg silk,

4. Suggested projects

(a) The testing of certain food for adulteration

1. Meat—for sulphites and borates
2. Milk—for formaldehyde
3. Butter—for oleomargerine, etc.

Horton, *Laboratory Manual in Chemistry.* pp. 63-66.

Ames and Jaffe, *Laboratory and Workbook Units in Chemistry*, pp. 178-180.

McGill and Bradbury, *Chemistry Guide*, pp. 285-288.

(b) The testing of textiles

1. Cotton
2. Wool
3. Silk
4. Rayon
5. Linnen.

Horton, *Laboratory Manual in Chemistry* pp. 72-79.

McGill and Bradbury, *Chemistry Guide* pp. 289-290.

(c) Determination of acetic acid in vinegar

Horton, *Laboratory Manual in Chemistry*, pp. 47-48.

McGill and Bradbury, *Chemistry Guide*, pp. 277-78.

NOTES

1. Brownlee et al., *First Principles of Chemistrv*, Allyn and Bacon, 1940, pp. 390-451; 661-700.
2. Howard, *Units in Chemistry*, Henry Holt, 1938, pp. 613-657.
3. Grer. and Bennett, *Chemistry*, Allyn and Bacon, 1939, pp. 382-686.
4. Jaffe, *New World of Chemistry*, Silver Burdett, 1940, pp. 349-371; 558-638.
5. Black and Conant, *New Practical Chemistry*, Mcmillan, 1936, pp. 217-239; 423-453; 560-574.
6. Brauer, *Chemistry and its Wonders*, American, 1938, pp. 441-551.
7. Horton, *Modern Everyday Chemistry*, Heath, 1937, pp. 169-219; 305-411.
8. Biddle and Bush, *Dynamic Chemistry*, Rand McNally, 1937, pp. 597-739.
9. Brownlee et al., *Chemistry in Use*, Allyn and Bncon, 1939, pp. 177-201; 396-476; 520.561.
10. McPherson, Henderson, and Fowler, *Chemistry at Work*, Ginn. 1938. pp. 137-162; 428-499.
11. Hopkins et al., *Chemistry and You*, Lyons and Carnahun. 1939, pp. 425-583.
12. Kruh, Carleton, and Carpenter, *Modern-Life Chemistry*, Lippincott, 1937, pp. 595-691.
13. Gray, Sandiphur, and Hannah, *Fundamental of Chemistry*, Houghton Mifflin, 1938, pp. 448-464; 477-582.
14. Bruce. *High School Chemistry*, World Book Company, 193, pp. 298-313; 491-515.

15. Dull, *Modern Chemistry*, Henry Holt, 1940, pp. 282-297; 678-745.

16. Wilson, *Descriptive Chemistry*, Henry Holt, 1936, pp. 58-110; 185-224; 224-443.

17. Wilson and Mullins, *Applied Chemistry*, Henry Holt, 1939 pp. 93-159; 245-291; 359-443.

CHEMISTRY

TEACHER EVALUATION AND SUGGESTIONS FOR IMPROVEMENT

Unit........... ....
(*Number*) (*Title*)

Note to Teacher: As a result of your experienecs in teaching this unit, please make suggestions for improvement and send to Lloyd W. King, State Superintendent of Public Schools. Use an additional sheet if necessary.

Specific Objectives

Subject Content

Learning Activities

Index